NORTH CAROLINA
STATE BOARD OF COMMUNITY COLLEGES
LIBRARIES
ASHEVILLE-BUNCOMBE TECHNICAL COMMUNITY COLLEGE

Construction safety handbook

Construction safety handbook

V. J. Davies, CEng, FICE
K. Tomasin, DLC(Hons)Eng, CEng, MICE, FIExpE

Thomas Telford, London

Published by Thomas Telford Ltd, Thomas Telford House, 1 Heron Quay, London E14 9XF.

First published 1990

Cover photograph: courtesy Dartford River Crossing Limited

British Library Cataloguing in Publication Data
Davies, V. J.
　Construction safety handbook
　1. Great Britain. Construction. Safety measures
　I. Title　II. Tomasin, K.
　624′.028′9

ISBN 0 7277 1385 X

© V. J. Davies and K. Tomasin, 1990

All rights, including translation, reserved. Except for fair copying, no part of this publication may be reproduced, stored in a retrieval system or transmitted in any form or by any means, electronic, mechanical, photocopying, recording or otherwise, without the prior written permission of the Publications Manager, Publications Division, Thomas Telford Ltd, Thomas Telford House, 1 Heron Quay, London E14 9XF.

Typeset in Great Britain by The Alden Press, Oxford
Printed and bound in Great Britain by Redwood Press Ltd, Melksham, Wilts

Acknowledgements

The authors wish to thank their colleagues in the Health and Safety Executive and W. S. Atkins Consultants for permission to use material in this book based on papers and other literature prepared in the course of their employment. They also wish to thank the British Standards Institution for permission to draw from contributions to several codes of practice and to the Health and Safety Executive for information from several of its publications.

The authors greatly appreciate the valuable help given them by John Lomax, OBE, Safety Consultant (chapters 8 and 9); Malcolm Tucker, John Laing Construction Ltd (chapter 5); Clive Tucker, British Rail (chapter 5); Anthony Cuming, W. S. Atkins Engineering Sciences (chapter 7); Tim Foley, W. S. Atkins Consultants (chapter 2, part 2); and Mr Tomasin's daughter Jill (chapter 2).

Contents

Introduction	1
1. The problem	3
2. The law	11
3. Enforcement of health and safety law	28
4. Hazards of construction and their prevention	40
5. Safety policies	140
6. Management systems for safe construction	162
7. Safety and reliability	191
8. Information	196
9. Training	204
10. Protective clothing and safety equipment	208
11. Accidents and first aid	218
References	232
Appendix 1. Acts and Statutory Instruments relevant to the construction industry	234
Appendix 2. Building Regulations 1985	235
Appendix 3. Health and Safety Executive addresses	236
Appendix 4. HSC/E publications relevant to construction	239

Appendix 5. Contents of first-aid boxes and travelling first-aid kits	246
Appendix 6. Dangerous occurrences to be notified (abridged)	247
Index	249

Introduction

Attention to matters of health and safety is a responsibility of everyone at work but it is of particular importance in the construction industry where the accident rate is so high. The initiative for this book developed from discussions on safety training by the Safety in Civil Engineering Committee of the Institution of Civil Engineers. Effective training of professionals in the construction industry is one of the means by which safety management can be improved and the number of injuries and fatalities reduced. It was recognised that there was a need for a textbook for graduate civil engineers wishing to study for their professional examinations and to give the qualified professional engineer, architect, surveyor and builder as part of his continuing education, sufficient information to

- take care for his own safety
- manage the safe conduct of those employees for whom he is responsible
- fulfil his moral and legal duty of care for the safety of others.

The book therefore has been written primarily by civil engineers for civil engineers but its contents are relevant for architects, surveyors, builders and all professionals and managers engaged in construction and demolition.

Responsibility for safety and health is not only confined to construction work on site. Design engineers, architects and surveyors are exposed to hazards during the investigatory stage of a project and while carrying out inspection tasks during the construction phase and on completed works. In addition, designers carry both a moral responsibility and a duty of care for the safety of building workers, maintenance staff, in some cases, demolition workers, and towards the public in general.

To avoid the repeated use of the word 'person' the authors have resorted to a number of alternative words and phrases that do not mean precisely the same. Where 'a workman might be injured' is used, 'a person might be injured' is meant. The persons will usually be workmen but not always — they might be professional staff, they might be visitors to the site, or even members of the public. The masculine gender has been used for convenience in several places but it is not to imply any discrimination on the part of the authors and we sincerely trust that the increasing number of female civil engineers, architects, builders and surveyors will not be offended.

1

The problem

Safety and health

Safety in the context of civil engineering is the discipline of preserving the health of those who build, operate, maintain and demolish engineering works and of others affected by those works. Safety is defined as the freedom from danger of risks. This can apply equally to the danger of physical injury and to the risk of damage to health over a period of time. However, in this book the term safety generally applies to the freedom from risk of injury from accidents whereas health damage arises from both immediate and longer-term effects of exposure to an unhealthy working environment.

Safety

Accidents that occur during construction and demolition activities result in injury, mostly, but not invariably, to employees on the site. Accidents can occur even before works begin, during survey and investigatory phases of a project, and they can also occur after the works have been completed, because of faulty design or construction, causing death or injury to those engaged on maintenance work and to members of the public.

The construction industry accident record

In a typical decade about 1500 people are killed on construction sites in Britain and 25 000–30 000 more are seriously injured. In addition, 300 000–400 000 suffer injuries sufficient to keep them off their normal work for at least three days.

Table 1.1 shows the annual deaths and major non-fatal injuries to employees in the construction industry for 1981–89. In 1985 the procedures for reporting injuries, diseases and dangerous occurrences (*see* p. 229) were revised which disturbed the continuity of major injury statistics. However, the injury rate in the construction industry continues at an unacceptably high level.

Unfortunately, it is not only the construction workers themselves who suffer injuries and death. In $7\frac{1}{4}$ years to March 1988 at least 88 people not employed in the industry, including more than 21 children, were killed on construction sites or because of construction activities, and a further 875 were seriously injured.

Table 1.1. Fatal and major injuries in construction

	1981	1982	1983	1984	1985	1986 Jan/Mar	1986/7	1987/8	1988/9
Deaths	128	135	150	130	138	26	125	143	136
Major injuries	1727	1999	2232	2356	2357	576	3179	3328	3620
Total	1855	2134	2382	2486	2495	6C2	3304	3471	3756

Note 1. From 1986 the reporting year is April to March.
2. Fatalities and major injuries to the employed and self-employed are included in the table but not the non-employed.
3. The continuity of major injury statistics was disturbed in 1986 by a revised definit on of a major injury. *See* p. 229.

Source: Health and Safety Executive.

Table 1.2. *Fatal and major injury rates per 100 000 employees*

	1981	1982	1983	1984	1985	1986/7	1987/7	1988/9
Fatal injury								
Construction industry	9.7	9.7	11.6	9.8	10.5	10.2	10.3	9.8
Manufacturing industry	2.0	2.4	2.2	2.7	2.4	2.1	1.9	1.8
Major injury								
Construction industry	155	188	213	225	225	282	276	282
Manufacturing industry	69	72	80	90	92	145	142	141

Note revised definition of major injury from 1986. *See* footnote to Table 1.1.
This table does not include the self-employed.

To put accident figures in perspective they must be linked to the number of people at risk. Thus, comparisons of incidence rates, which are the number of fatalities and major injuries per 100 000 employed, provide a means of assessing the relative danger for people at work in various industries. Table 1.2 compares the injury rates of the manufacturing industry with those of the construction industry for 1981 to 1988–89. In the construction industry the risk of major injury is two to two-and-a-half times greater and the risk of a fatal accident nearly five times greater than in the manufacturing industry.

Economic effect of accidents. The accident statistics represent not only terrible human tragedies but also substantial economic cost because accidents also cause

- damage to plant and equipment
- damage to work already completed
- loss of productive work time while debris is cleared and damaged work rebuilt
- reduced workrate until normal site working rhythm and morale are restored
- disruption while investigations are carried out by the company safety department, the insurers, inspectors from the Health and Safety Executive and sometimes representatives from the trade unions
- legal costs and, in some cases, fines
- increased insurance premiums
- loss of confidence and reputation.

All this has to be balanced against the cost of working safely. Insurers can show that the modest extra cost of a safe, orderly site is well invested because the risk and cost of serious accidents is so high.

Reasons for the poor record. There are a number of reasons why the accident record in the construction industry compares poorly with that of the manufacturing industry.

In factories there is normally a controlled working environment, with little change in working procedures and equipment for long periods, and the labour force usually remains fairly constant. Hazards once identified can be remedied with relative ease and the danger can be overcome: for example, a guard may be designed and fitted to an unfenced part of dangerous machinery.

However, in the construction industry the working environment is constantly changing, sites exist for a relatively short time and the activities and inherent risks change daily. Within a short time of a hazard being identified and dealt with, the work scene has changed, bringing new hazards. There is also a high turnover in the workforce, which means safety awareness is not always as good as it should be.

The complexity of the construction industry with its many small firms, subcontractors and self-employed labour militates against the establishment of safe working practices. The financial resources of small firms are generally insufficient to provide the necessary high standard of safety training carried out by many of the larger firms. Moreover, the managerial knowledge of many small contractors is based on experience and often lacking in theoretical background. Many small contractors do not employ a professional safety adviser and have neither the time nor the inclination to keep abreast of legal requirements and technical developments in safety and health matters. Compared with accidents in factories there are many more accidents in the construction industry that cause multiple fatalities and serious injuries.

When competition is particularly fierce, some contractors are prepared to ignore safety legislation and good working practice in order to bid competitively. Unfortunately, there are also clients who are willing to employ these contractors regardless of their competence in safety management.

Safety legislation and, in particular, the Construction Regulations (*see* p. 13) provide a sound basis for good safe building practice. However, the Factory Inspectorate of the Health and Safety Executive, whose duty it is to enforce the law, does not have the resources to supervise every site. Unfortunately, therefore, law breaking is rife because the risk of prosecution is low and, moreover, the practice of cutting corners for commercial gain is frequently condoned by senior management and ignored by clients and their professional advisers.

The majority of accidents can be prevented by dedicated safety management and a disciplined approach to ensure compliance with safety rules and procedures by all involved in construction operations. This should be evident from the top management down. The tough image which flouts all foreseeable dangers is a major obstacle to be overcome for it not only causes individual casualties but also affects large sections of the industry, resulting in the widespread axiom that, 'if you are not prepared to take risks you should not be in the business'.

Causes of accidents

The key to influencing safety and health is to foresee the hazards and thus be in a position to eliminate them. The main causes of accidents are

- falls
- stepping on or striking against objects
- lifting and carrying — over-exertion
- machinery
- electricity
- transport
- fires and explosions.

By far the largest category is falls, which includes people falling from one level to another, people falling at the same level and plant and material falling (including a structure or part of a structure collapsing) and striking, crushing or burying people. Each year 70–80% of all fatalities and 35–40% of all injuries may be attributable to this cause.

About 25% of all the accidents that result in workers being unable to carry out their normal work for at least three days are due to incorrect

8 CONSTRUCTION SAFETY HANDBOOK

Table 1.3. Causes of fatal injuries in the construction industry for 1981–85

Falls (of people)	52%
Falling material/objects	19%
Transport/mobile plant	18%
Electricity	5%
Asphyxiation/drowning	3%
Fire/explosions	2%
Miscellaneous	1%

manual lifting and carrying, generally of too-heavy loads. This usually results in strain and sprain injuries. A further 10% are due to stepping on or striking against objects, for example, stepping on protruding nails left in discarded timber. Fortunately, few accidents in these categories are fatal.

Fatalities. An analysis of fatal accidents in the building and civil en-

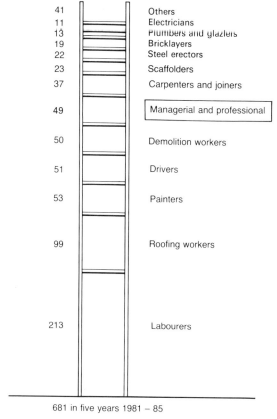

681 in five years 1981 – 85

Fig. 1.1. HM Factory Inspectorate's distribution of fatalities by occupation for 1981–85 (courtesy HSE; source Blackspot Construction)

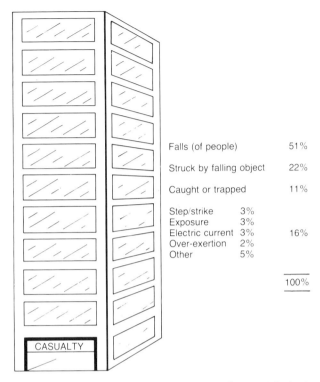

Fig. 1.2. *The construction industry hospital: causes of major injuries in a typical year (compiled from HSE sources)*

gineering industries published by the Health and Safety Executive in 1988 entitled *Blackspot Construction* describes how people die.[1] Of the 681 deaths in five years 1981–85, the causes were as shown in Table 1.3.

Figure 1.1 illustrates HM Factory Inspectorate's distribution of fatalities by occupation for 1981–85. It takes no account of the numbers employed in the various occupations and trades and should, therefore, not be regarded as an indicator of the relative risk attached to each group. However, professional engineers, together with managerial staff, suffered 49 fatalities — some 7% of the total — during the period.

Major injuries. Specified major injuries are defined by the Health and Safety Executive for reporting purposes and are listed in chapter 11 (on p. 229). The construction industry hospital (Fig. 1.2) illustrates the approximate distribution of all major injuries that occur in a typical year, from which it can be seen that 73% were due to people, materials or objects falling.

Reliability of the statistics

While the number of fatalities and specified major injuries at work reported to HM Factory Inspectorate can be expected to be reasonably accurate, there is reason to suspect under-reporting of the other over-three-day injuries required by RIDDOR (*see* p. 229). The reasons for this are that some firms may not comply with reporting procedures, either because of ignorance or from unwillingness to disclose accidents, and also that the growth of sub-contracting and self-employed labour on construction sites has tended to confuse the responsibility for reporting. In addition, some people who have suffered injuries that keep them off normal work for more than three days continue work on light duties, and their injury is often wrongly not reported. It is likely therefore that the accident rate in the industry is somewhat worse than the statistics show.

Health
Occupational ill-health statistics
In comparison with accident statistics, the recording of cases of occupational diseases is beset with problems of two sorts

- defining what is meant by occupationally related ill health
- devising methods reliably to count (or estimate) those occurrences of ill health falling within the definition.

Nevertheless, the Health and Safety Executive publish statistics based on cases of notifiable diseases reported to them and from other sources. Many cases of occupational ill health occur to construction workers, but separate statistics for the industry are not available. The procedure for reporting diseases is described on p. 230.

Health hazards. The major health problems associated with construction work may be categorised into

- chemical hazards: those due to liquids, dusts, gas and fumes
- physical hazards: those due to cold, heat, noise vibration, ionising and non-ionising radiations, and compressed air.

For works in sewers and on contaminated sites there is also a risk of biological infection. The common health hazards of construction operations are summarised on p. 131 *et seq.*

Employers are required by law (*see* p. 20) to identify health hazards associated with the work of their employees and to take the necessary steps to remove the hazards or to give adequate protection both to their employees and anyone else who may be affected.

2

The law

Introduction

An examination of safety and health in the construction industry would be unrealistic and inadequate without at least a brief consideration of the law. A detailed analysis of all the legal problems likely to affect engineers wherever they are on the management ladder is not given here, but engineers should be aware of the basic nature of their duties and liabilities both in criminal and civil law. In general terms there are two main branches of English law distinguished not in the nature of the act someone commits (for the same act can constitute both a crime and a civil wrong) but in the legal consequences that may follow from that act. The terminology relating to both branches of the law differs and the criminal and civil courts are also almost entirely distinct.

PART 1. CRIMINAL LAW
Basis of criminal law

The law relating to health and safety at work is predominantly criminal law enforced through proceedings brought in the magistrates and crown courts. On conviction, any employee or his company may be fined a sum which is paid to the State, or in extreme cases he (or an appropriate officer of the company) may be sentenced to a term of imprisonment.

The most important form of legislation is the Act of Parliament and much of health and safety law is embodied in Statute. The Health and Safety at Work etc. Act 1974 (HSW Act) is an example of such an Act of Parliament. There is, in addition, delegated or subordinate legislation; Government Orders known as Statutory Instruments, which often contain detailed rules and regulations relating to Acts of Parliament. The Construction Regulations of 1961 and 1966 made under the Factories Act 1961 are examples of delegated legislation.

History of construction health and safety legislation

As long ago as 1904 concern was shown over accidents and working conditions in building work. A Public Enquiry was held, but it was not until 1926 that the first Building Regulations were passed by Parliament. However, they were of limited scope and only applied to sites on which mechanical power was used. In 1931, their scope was widened to cover the growing number of cranes used in building work, the main types of which were scotch derricks and guyed derricks, where the anchorage and ballasting arrangements were often inadequate.

By the late 1930s these regulations had become outdated but because of the intervention of the Second World War, it was not until 1948 that the first comprehensive set of Building Regulations were compiled. They applied to all building sites (but not to civil engineering works) where persons were employed. These became law following a Public Enquiry and were entitled The Building (Safety Health and Welfare) Regulations 1948.

The post-war expansion of the construction industry, coupled with an equally large increase in the accident rate and the need to improve safety in the civil engineering section of the industry, called for more complex and comprehensive legislation. This resulted in the replacement of the 1948 Building Regulations by four Statutory Instruments collectively known as the Construction Regulations. The first two, the Construction (General Provisions) Regulations 1961 and the Construction (Lifting Operations) Regulations 1961 came into operation in 1962. Civil engineering works were then for the first time covered by these legal requirements. They were followed in 1966 by the Construction (Working Places) Regulations 1966 and the Construction (Health and Welfare) Regulations 1966. The Construction Regulations of 1961 and 1966 were Statutory Instruments made by various Ministers of Labour through powers conferred on them by The Factories Act 1937, 1948, 1959 and 1961.

In 1974, a major piece of legislation called the Health and Safety at Work Act 1974 etc. was introduced. This Act, the Factories Act 1961 and the Construction Regulations are described in the following paragraphs. A list of Acts of Parliament, Statutory Instruments etc. relevant to the construction industry is given in Appendix 1.

The Factories Act

The Factories Act 1961 is the principal Act concerned with the safety, health and welfare of people employed in factories, although the provisions have been extended to places other than factories. Section 127 of the Factories Act stipulates the parts which apply to building operations and work of engineering construction (civil engineering work). Also, under this section, building and civil engineering operations are subject to section 76, which provides the powers for the Secretary of State for Employment to make special regulations for safety and health. Other provisions under section 127 extend the powers and duties of HM Inspectors of Factories, to cover construction work, e.g. to visit and inspect construction sites. Section 127 requires any person undertaking construction work to which the Act applies to keep certain records and registers, and to make them available for examination by HM Inspectors of Factories.

Another important requirement of section 127 of the Factories Act is that any person undertaking construction work to which the Act applies must inform, in writing, the District Inspector of Factories (from 1975, the Principal Inspector of the Health and Safety Executive) responsible for construction activities where the work is taking place, the name and address of the firm or persons undertaking the work, the place and nature of the work and other prescribed particulars. This must be done within seven days of the work starting. Notification is not required if the work will take less than six weeks to complete (although there are proposals to reduce this period) or if the particulars have been furnished by others, for example, where a main contractor has covered his subcontractors in his return.

The Construction Regulations
The Construction Regulations of 1961 and 1966 comprise

- The Construction (General Provisions) Regulations 1961
- The Construction (Lifting Operations) Regulations 1961
- The Construction (Working Places) Regulations 1966
- The Construction (Health and Welfare) Regulations 1966.

Although the four sets of regulations are written in legal terminology the requirements are easy to understand. Nevertheless simplified versions and guides are published by various bodies such as the Federation of Civil Engineering Contractors, the Building Employers Confederation and the Royal Society for the Prevention of Accidents (*see* p. 202).

The Construction (General Provisions) Regulations 1961. The regulations contain provisions dealing with health and safety in specialised operations such as excavations and tunnelling work, shaft, cofferdam and caisson construction, road and rail transport, work on or adjacent to water, the use of explosives, dangerous or unhealthy atmospheres, and demolition work. Miscellaneous items, for example, fencing of machinery, electricity, lighting, site tidiness, protection of eyes, protection from falling material, construction of temporary structures and the avoidance of collapse of any part of a structure during construction, are also dealt with.

One requirement is that every contractor or employer who employs more than 20 people on construction work, not necessarily on the same site or at work at the same time, must specifically appoint, in writing, a safety supervisor (or more than one), suitably qualified, to advise him on the observance of the requirements imposed by or under The Factories Act for the safety and health of people carrying out construction operations. Also, the safety supervisors are required to exercise general supervision to ensure observance of those safety and health requirements on site, and to promote the safe conduct of the work generally. In meeting this obligation a contractor may appoint a safety supervisor to cover a number of sites, provided of course that these are not too many or too widely scattered to prevent him from properly carrying out the duty. It is acceptable for two or more contractors jointly to appoint the same person or persons. This is often done by smaller contractors who employ a firm specialising in advising on safety and health matters.

The Construction (Lifting Operations) Regulations 1961. The regulations deal with the construction, erection, inspection, examination and use of lifting appliances and lifting gear (tackle) on construction sites. They cover such plant and equipment as cranes, hoists, winches, piling frames, shear legs, excavators, draglines, pulley blocks, overhead runways, cableways, slings, shackles, eyebolts, hooks, wire and fibre ropes.

The regulations require that the plant and equipment be kept in good order and safe for use, and that it is used safely. They cover brakes, controls, safety devices, stability and anchorage arrangements of cranes and other lifting appliances. The testing and examination of cranes and winches etc. are covered in detail, as are the requirements for marking safe working loads, the provision of automatic safe load indicators and radius load indicators for cranes. The construction, examination, testing and safe working load for chains, ropes and lifting gear are also dealt with. The regulations also cover hoists and the carriage of people by lifting appliances.

The Construction (Working Places) Regulations 1966. The regulations are designed to ensure the provision of a safe means of access to and egress from every place where a person has to work. Also that the place of work itself is made and kept safe.

The regulations contain provisions dealing with scaffolding, ladders, cradles, safety nets and belts. Methods of working in dangerous places such as on sloping roofs, near openings and edges, over open joisting and in the vicinity of roofs constructed of fragile materials are also dealt with, together with the inspection of scaffolds and the formal reporting of the inspections.

Detailed requirements include the minimum dimensions for working platforms, gangways and runs, and the provision of guard rails and toe boards. Construction requirements for scaffolds, both at interim and completed stages, and the quality of the material to be used, are also covered.

One important regulation concerns scaffolds used by workers of different employers. It requires that before an employer allows his employees to use a scaffold he must satisfy himself either personally or through a competent agent that it is safe for use and complies with the regulations. No matter what is written into or implied by the contract documents about the provision of scaffolding, this regulation is applicable. Thus it is the duty of all employers, including subcontractors, clients, consulting engineers and architects, whose employees use scaffolding erected by the main contractor, to ensure that the scaffolding is safe for their employees.

The Construction (Health and Welfare) Regulations 1966. The regulations are concerned with the welfare of employees rather than with their safety. They cover the provision of first-aid equipment, washing facilities, sanitary conveniences, shelters and accommodation for clothing and taking meals, and protective clothing where work is required to continue outside during inclement weather.

The regulations have since been amended. The first-aid provisions have been revoked and new requirements to replace these included in the Health and Safety (First Aid) Regulations 1981. The welfare part of the regulations, however, remains unchanged.

Obligations under the regulations. An important clause that appears in each of the four sets of Construction Regulations is headed Obligations under the Regulations, commonly referred to as the duties clause because it indicates specifically who has a duty to comply with particular requirements.

Every contractor and employer has a duty under the regulations to provide for the safety, health and welfare of his own employees whether or not the hazard has been created by someone else. However, where an employer's work is of such a nature that employees other than his own may be endangered, for example, in roof work, in the use of explosives or from demolition operations, he has a duty to safeguard all affected workers whether employed by him or not.

Similarly, those who erect, install or use plant such as lifting machinery and gear are required to do so in accordance with the Construction (Lifting Operations) Regulations 1961, and so the user as well as the erector must comply with the regulations. Where hired cranes are used, the contractor directing and using the machinery has duties to comply with the regulations even though the driver may be employed by the hire firm.

Employees also have duties under the obligations clause to comply with certain regulations that apply to them. Where, for example, a regulation requires the use of a certain piece of equipment, if a contractor supplies it for use and the employee does not use it, the employee may well be in breach of his statutory duty.

If a duty is owed by a contractor/employer under the obligations clause, the obligation cannot be passed on to others by a clause in the contract documents or by any other means. There is, however, an exception to this in the Construction (Health and Welfare) Regulations 1966. The duties relating to the provision of shelters, messrooms, washing facilities and sanitary facilities may be satisfied by making arrangements for sharing the facilities with another contractor on site or with any other person or body. This arrangement is only permissible if the facilities are adequate and conveniently situated for sharing purposes. The providing contractor is responsible for complying with the regulations in respect of all those who use the facilities.

The Health and Safety at Work etc. Act 1974

The Health and Safety at Work etc. Act 1974 (HSW Act) covers all people at work whether employers, employees or self-employed, except for domestic servants in private households and transport workers engaged in transport operations, for example, train drivers on British Rail and bus drivers. It also protects the general public where their health and safety may be affected by the work activities of others, for example, by contractors.

The HSW Act applies to England and Wales and, except for most of part III, to Scotland. Although the HSW Act does not apply to Northern Ireland, the duties imposed on employers and employees by the Health and Safety at Work (Northern Ireland) Order 1978 are substantially the same as for the HSW Act. The HSW Act is in four parts. Part I is principally concerned with safety in the construction industry.

HSW Act part I. The purpose of part I of the Act is to
- secure the health, safety and welfare of people at work
- protect people other than those at work against risks to their health or safety arising out of the activities of people at work
- control the keeping and use of explosives or highly flammable or otherwise dangerous substances and generally prevent people from acquiring and using such substances unlawfully
- control the release into the atmosphere of noxious or offensive substances from premises prescribed by regulations.

The general duties of employers and employees are covered in sections 2–9 of part I and are summarised as follows.

Under section 2 every employer has a duty to ensure the health, safety and welfare at work of all his employees, particularly by

- the provision and maintenance of plant and systems of work that are safe and without risk to health
- making arrangements for ensuring safety and absence of risk to health in connection with the use, handling, storage and transport of articles and substances
- the provision of the necessary information, instruction, training and supervision
- the maintenance of workplaces under his control in a safe condition and without risk to health; also the provision and maintenance of means of access to and egress from places of work that are safe and without risk to health
- the provision and maintenance of working environments which are safe, without risk to health and adequate as regards facilities and arrangements for their welfare.

Also, under section 2 all employers with five or more employees must prepare a written statement of their general policy with respect to the health and safety at work of their employees. The organisation and arrangements for carrying out the policy must be included and it must be revised as often as necessary to keep it up to date. The policy statement has to be brought to the attention of all employees.

Furthermore, recognised trade unions may appoint safety representatives from among the employees, and where they have been appointed, employers must consult them with a view to making and maintaining arrangements for ensuring the health and safety at work of all employees. Also, if requested in writing to do so, the employer must set up a safety committee whose function is to keep under review the measures taken to ensure the health and safety of employees while at work.

Section 3 requires every employer to conduct his undertaking in such a way as to ensure that people not in his employment are not exposed to risks to their health and safety. Similarly, every self-employed person must also conduct his undertaking so that he and other people (not being his employees) are not exposed to risks to their health and safety. Thus everyone, including members of the public who might be affected by the work activities, is protected under this section.

Section 4 imposes duties on those who to any extent have control over premises (other than domestic premises), the means of access, or of any plant or substances in the premises, which are used by people who are not their employees as a place of work. They must take measures to ensure that they are safe and without risk to health.

Section 5 requires those who control premises to use the best practical means for preventing the emission of noxious or offensive substances into the atmosphere and for rendering harmless and inoffensive any that do escape into the atmosphere.

Section 6 places duties on persons who design, manufacture, import or supply (by sale or hire) articles for use at work. They must

- ensure that the article is designed and constructed in such a way that it is safe and without risk to health when properly used
- make arrangements for any testing and examination that may be necessary; designers and manufacturers are also required to carry out or arrange for the carrying out of any necessary research with a view to the discovery, elimination or minimisation of any risk to health and safety to which the design or articles may give rise
- take such steps as are necessary to ensure that adequate information will be available about the use for which the article has been designed and tested
- ensure, if they erect or install the article in any premises where it is to be used by persons at work, that it is installed or erected so as to be safe and not a risk to health when properly used.

An article is not regarded as properly used where it is used without regard to any relevant information or advice relating to its use which has been made available by a person by whom it was designed, manufactured, imported or supplied. Similar requirements apply to the supply etc. of substances for use at work.

Section 7 places duties on employees while at work. They must

- take reasonable care for the health and safety of themselves and of others who may be affected by their actions or omissions at work, and
- co-operate with their employers or anyone else as necessary to enable them to comply with their statutory duties.

Section 8 places a duty on all people, both those at work and members of the public, including children, not intentionally to interfere with or to misuse anything that has been provided in the interests of health, safety and welfare.

Section 9 prohibits an employer from charging his employees for anything done or provided for health and safety purposes to comply with a specific statutory requirement.

The remaining 45 sections of part I of the HSW Act are concerned with the powers and duties of the Health and Safety Commission and Executive, enforcement, offences and miscellaneous matters (*see also* chapter 3). However, special mention is made here of the issue of approved codes of practice.

Approved codes of practice. Under sections 16 and 17 of the HSW Act the Health and Safety Commission may approve and issue codes of practice, whether prepared by it or by some other body, for example, British Standards Institution for the purpose of providing practical guidance on the requirements of section 2–7 of the HSW Act or of any health and safety regulations or statutory provisions.

Failure to observe the guidance provided by an approved code of practice is not in itself unlawful. However, if someone is prosecuted for a breach of any requirement of sections 2–7 of the HSW Act or related legislation and an approved code of practice is relevant to that breach, the code is admissible as evidence. Further, it is then up to the defendant to show either that the code was complied with or that what was done was as good or better than the provisions in the code.

By the end of 1989, 40 codes had been designated as approved, 13 of which are relevant to the construction industry. They cover such subjects as safety representatives and safety committees, the training of safety representatives, control of lead at work, work with asbestos insulation, asbestos coating and asbestos insulation board, the Health and Safety (First Aid) Regulations 1981, packaging of dangerous substances for conveyance by road, protection of people against ionising radiation arising from any work activity, exposure to radon and safety in docks. A complete list of current approved codes of practice may be obtained from the Health and Safety Executive (*see* p. 200 and Appendix 4).

HSW Act Part II. Part II of the HSW Act describes the functions of the Employment Medical Advisory Service and its responsibility for, among other things, giving employed people and those seeking employment, information and advice on health at work.

HSW Act Part III. Part III is concerned with the making of Building Regulations, which is the responsibility of the Department of the Environment and the Scottish Office. The majority of part III has since been deleted from HSW Act and included in the Building Act 1984.

The Building Regulations are for securing the health, safety, welfare and convenience of those in or about buildings. They are principally concerned with the design and construction of buildings and the provision of services, fittings and equipment in them. They are therefore of primary relevance to building owners and designers. The subjects they cover are listed in Appendix 2.

The Construction Regulations, however, are concerned with the safety, health and welfare of those engaged in the construction, maintenance and demolition of works.

HSW Act Part IV. Part IV contains various miscellaneous and general provisions, for example, amendments to the Radiological Protection Act 1970, amendments to the Fire Precautions Act 1971 and powers to repeal or modify Acts and Instruments and the extent and applications of the HSW Act.

Construction Management Regulations (Proposed)

A consultative document was issued by the Health and Safety Commission in September 1989 seeking comments on proposals for a new set of regulations, together with a supporting approved code of practice. These may be known as the Construction Management (and Miscellaneous Duties) Regulations, aimed at strengthening the legal requirements relating to the management and control of health and safety on construction sites.

It is intended to place duties not only on contractors and subcontractors but also on clients and designers. It is proposed that 'every person who has, to any extent, control over another person who is at work on a construction site shall make and take such measures as is reasonable for a person in his position to take', to carry out adequate arrangements for the management of the site to ensure the safety of anyone who may be affected by the construction works.

It is proposed that the client shall take reasonable measures to ensure that those with whom he enters into a contract, whether as a contractor or provider of professional services, has the competence necessary to perform the contract without risk to health and safety. Clients must also ensure that contractors are provided with all the information about the state or condition of the land or premises which might affect the health and safety of those affected by the works.

Designers will have two main duties. First to make sure that the designed structure can be built, maintained, repaired and demolished safely and, second, to ensure that adequate information about the design or materials that might affect the health or safety of any person is passed to the contractor.

Interpretation of the term 'reasonably practicable'. Many of the duties in the HSW Act are qualified by the term 'so far as is reasonably practicable'. For example, 2(1) states that 'it shall be the duty of every employer to ensure *so far as is reasonably practicable* the health, safety and welfare at work of all his employees'. This term is not defined in the Act, but the courts have interpreted it as requiring an assessment to be made of the degree of risk which has to be weighed against the cost, in money, time and trouble of averting the risk. If the cost is high compared with the likelihood and severity of an injury (or ill health) being sustained it may well be considered not reasonably practicable to comply with the requirement. However, if the likelihood of a severe injury (or ill health) being sustained is high when a certain method of work is employed then it may be considered reasonably practicable to require considerable expense to be incurred to avert the likelihood of such injury, i.e. to remove the hazard.

Whenever the expression is used without the qualification of 'reasonable', i.e. 'so far as is practicable', it is a much stricter standard and, if it is possible to obviate the hazard, then it must be done at whatever cost.

Where proceedings are instituted under the HSW Act alleging failure to comply with a requirement of duty to do something 'so far as is practicable' or 'so far as is reasonably practicable' or 'to use the best practicable means', by virtue of section 40, it is for the accused to prove that it was

Substances hazardous to health

Regulations known as the Control of Substances Hazardous to Health Regulations 1988 came into force on 1 October 1989. The aim of the regulations, made under section 16 of the HSW Act, is to maintain and improve existing standards of health and safety concerning the protection of people against risk to their health, whether immediate or delayed, arising from exposure to hazardous substances. A number of sections of existing Acts and about 50 sets of regulations and orders are repealed by the new regulations.

The regulations apply to substances that are classified as toxic, harmful, corrosive or irritant and to those which have maximum exposure limits and those with occupational exposure standards (OES). They also cover substances that have chronic or delayed effects. The meaning of substances hazardous to health is given in regulation 2.

It is the duty of every employer not to expose employees to hazardous substances unless he has made an assessment of

- the risks to health created by the work
- the steps to be taken to comply with the regulations.

Every employer must

- prevent or control exposure
- use, maintain, examine and test the control measures
- monitor exposure at the workplace
- carry out health surveillance
- provide information instruction and training.

The duty imposed upon the owner or occupier of the premises is to carry out the above measures for the protection not only of his own employees but also of others at the premises or likely to be affected 'so far as is reasonably practicable', except that he is not responsible for health surveillance of employees not his own. This is specially relevant to civil engineers and others in the construction industry. For employees working on sites or establishments where there are hazardous substances the onus is upon the employer to provide the required health surveillance of his employees. Duties are also imposed upon the self-employed, except for monitoring exposure and health surveillance.

Test of the HSW Act in the courts — HMS Glasow

A significant case in the administration of health and safety legislation arose from an incident during the construction of HMS *Glasgow* at Swan Hunter Shipbuilders: *R.* v. *Swan Hunter Shipbuilders and another* [1982] 1 All E.R. 264. Subcontractors who had no contractual relationship with the shipbuilders were working on the ship while it was being fitted out. The shipbuilders were aware that because of the use of oxygen there was a danger of fire in confined spaces but did not distribute their rule book to subcontractors. A subcontractor's employee failed to disconnect an oxygen hose and a serious fire resulted.

The shipbuilders were prosecuted and found guilty of

- failing to provide safe systems of work
- failing to provide information and training
- failing to preserve the safety of employees not their own.

Following their appeal it was held that

- The duties imposed on an employer by the HSW Act followed the common law duty of care of a main contractor to co-ordinate operations to ensure the safety of his own and other employees. The main contractor's protection lay only in the provision of the duties insofar as it was reasonably practicable to perform them. It was for the main contractor to prove that it was not reasonably practicable.
- It followed that, if it was reasonably practicable to give his own employees information and training then it was also reasonably practicable to give it to subcontractors' employees, both for the safety of his own employees and for the safety of subcontractors' employees and others.

PART 2. CIVIL LAW
Basis of civil law

Contractual disagreements and compensation matters are examples of civil law which is based on the principle that like cases should be decided alike. In applying the law to a given set of facts, the court refers to the decisions of judges in decided cases, applying the doctrine of precedent. If liability is proven, in most cases damages are paid to the injured party. Civil law is mainly made up of

- the law of contract
- tort.

A contract is based on an agreement between two or more parties. The law of contract exists to protect the agreements, to ensure that the parties to them carry out the promises they have made and to provide one party with a remedy should the other fail to do what he has undertaken. The remedy for breach of contract may be liquidated damages if the parties have agreed to a pre-estimated sum to be paid in the event of a breach. If, however, as is often the case, the contract is silent regarding any sum to be paid in the event of a breach then the damages to be awarded are left to a decision of the court. This is known as unliquidated damages.

Tort, however, has nothing to do with agreement. Tort is a wrong, a breach of duty towards persons generally, such as nuisance or negligence for which the remedy is an action for unliquidated damages.

In the context of construction safety it will be necessary to consider both contract law and tort and, as will be seen, it is likely that in many cases overlaps will occur involving both forms of civil law and possibly criminal law as well.

The law of contract

Contract is the basis of all commercial law and a contract is usually made by the acceptance by one person of an offer made by another. The agreement may be arrived at in writing, by spoken word or merely by conduct or a combination of all three.

For a legal contract to exist, a number of conditions must be satisfied, one of which is that the performance of it must be legal. A contract for carrying out construction works is possibly voidable if the performance of it requires the rules of common law or a statutory provision, for example, the Construction Regulations, to be contravened, and it follows that tenders must cover for complying with the law, for example, the HSW Act.

Conditions of contract

For most building and civil engineering contracts, it is necessary to define the conditions under which the contractor will be paid for the work he performs. They are usually known as the conditions of contract. Because of the similarities between various types of work, a number of standard conditions of contract are prepared by trade federations, institutions and some major authorities and commercial undertakings. The ones with which most civil engineers will be familiar are the ICE Conditions of Contract. In the context of safety on site, other conditions of contract may be more or less explicit but, whatever is stated in any contract document, the provisions of criminal law cannot be circumvented or delegated.

ICE Conditions of Contract

The Conditions of Contract for use in connection with Works of Civil Engineering Construction, commonly known as the ICE Conditions of Contract, are approved by and may be obtained from the Institution of Civil Engineers, the Association of Consulting Engineers or the Federation of Civil Engineering Contractors. The Conditions are commonly employed in the civil engineering sector of the construction industry in the United Kingdom and, in a modified form, in many overseas territories. The two parties to the ICE Conditions of Contract are the Employer and the Contractor. An Engineer is appointed by the Employer, and his representative (the Engineer's representative or, more popularly, the resident engineer) watches and supervises the works. The Engineer may be an independent consulting engineer or a member of staff of the Employer's organisation.

Clauses concerning safety. The most significant clauses in the fifth edition of the ICE Conditions of Contract that relate to site safety are clause 8(2), clause 19 and clause 40.

Clause 8(2) states

> The Contractor shall take full responsibility for the adequacy, stability and safety of all site operations and methods of construction provided that the Contractor shall not be responsible for the design or specification of the Permanent Works (except as may be expressly provided in the Contract) or of any Temporary Works designed by the Engineer.

Clause 19(1) states

> The Contractor shall throughout the progress of the Works have full regard for the safety of all persons entitled to be on the Site and shall keep the Site (so far as the same is under his control) and the Works (so far as the same are not completed or occupied by the Employer) in an orderly state appropriate to the avoidance of danger to such persons and shall *inter alia* in connection with the Works provide and maintain at his own cost all lights, guards, fencing, warning signs and watching, when and where necessary or required by the Engineer or by any competent statutory or other authority for the protection of the Works, or for the safety and convenience of the public or others.

It is therefore made quite clear that the contractor must construct the works in a safe manner as a condition of his contract with the Employer.

In another sub-clause, however, the Employer has responsibilities for safety if he has workmen or engages other contractors on the site. Clause 19(2) states

> If under Clause 31 the Employer shall carry out work on the Site with his own workmen, he shall in respect of such work:
>
> (a) have full regard for the safety of all persons entitled to be on the Site; and
> (b) keep the Site in an orderly state appropriate to the avoidance of danger to such persons.
>
> If under Clause 31 the Employer shall employ other contractors on the Site, he shall require them to have the same regard for safety and avoidance of danger.

Clause 31 cited in the above extract is concerned with facilities for other contractors on the site.

In the fifth edition of the ICE Conditions of Contract the responsibility of the Engineer for safety is not so clear. Under clause 40, the Engineer may suspend the progress of the works or any part of them if necessary for the safety of the Works. But there is no specific mention of the safety of employees or other people on or near the site. This subject is worthy of further discussion.

The ICE Conditions of Contract and the HSW Act. A number of editions of the ICE Conditions of Contract have been published, the first in 1945 and the fifth in 1973, one year before the HSW Act (1974). As a result, there are some aspects of the fifth edition that are not entirely compatible with current safety legislation. One of these is the role of the Engineer and his staff with respect to the safety of people on the site. The Engineer as an employer of engineers and inspectors has the primary responsibility for their safety under section 2 of the HSW Act. The engineer's representative may also be liable in certain circumstances under section 3 of the HSW Act for injuries or deaths arising from construction site accidents, particularly if people other than construction workers are among the casualties (*see* Fig. 6.1).

The primary purpose of the Conditions of Contract is to define the terms under which payment will be made and it is the financial risk of accidents and the delays resulting from them that the Contractor accepts by signing the Contract. However, the Contractor and everyone else on site must comply with criminal law and it follows that any event such as the serving of prohibition notice or an accident that results in the prosecution of the Contractor, or one of his workforce, is totally outside the Contract and any consequential costs must be borne by the Contractor. However, the cost incurred by the Contractor of any accident in which he is not to blame, would have to be determined by the Engineer and paid by the Employer.

Engineering contracts

Contracts to carry out the design, specification and supervision of civil engineering work range from a simple exchange of letters to legally prepared conditions of engagement signed by both parties on an offer and acceptance basis. The Association of Consulting Engineers (ACE) has drawn up standard Conditions of Engagement to form the basis of the agreement between the client and the consulting engineer for five different types of employment, including report and advisory work, civil, mechanical and electrical and structural engineering work, and engineering services. Section 5 of these conditions is concerned with care and diligence

> 5.1 The Consulting Engineer shall exercise all reasonable skill, care and diligence in the discharge of the services agreed to be performed by him etc.

Even if the ACE Conditions of Engagement are not used and the terms of appointment do not specifically mention that the engineer should use proper skill and care, the law will imply that he is bound to do so.

The question of professional skill and care has been considered by the High Court in *Greaves and Co. (Contractors) Ltd* v. *Baynham Meikle and Partners* (1975) 1 W.L.R. 1095. In this case, professional design services had been provided to the contractor to assist in a design-and-build project. The first floor failed because of insufficient attention to vibration effects of live loading on the structure and the consulting engineers were held to be liable for breach of contract. This decision is important to all professional engineers since it confirms that, if there are special factors that are known or ought to be known to a professional man and he is warned, perhaps in a code of practice, about dangers associated with them, then they constitute part of the ordinary skill expected of a competent engineer. In the special case of design-and-build contracts there is an obligation not only to design a structure that is safe, but also it must be reasonably fit for the purpose for which it is required unless the designer's obligation is expressly limited to exercising reasonable skill, care and diligence.

Limitation

The period for which a contractor or consulting engineer remains liable under the law of contract is six years in the case of an action on a simple contract and twelve years for a contract under seal. The time runs from the date when the cause of action accrues. In the case of contract the cause of action accrues at the moment of the breach of contract and it is irrelevant that damage is not suffered until a later date.

The limitations to the period of contract do not apply in the case of a fraudulent action. In the case of *Clark* v. *Woor* (1965) 1 W.L.R. 650, the defendant, a builder, contracted to build a bungalow with special bricks. In fact he used bricks that were inferior to those specified and eight years later in 1961 the plaintiff noticed some flaking of the bricks and discovered the breach of contract. The judge found that the plaintiffs relied on the builder, who was experienced, for honest performance of his contract. The builder had committed a fraud and consequently the plaintiff's action was not barred by the eight-year interval. The damages were calculated at prices in 1961, not in 1965 when the action was heard.

Failures involving injury at work
In the case of a site accident or failure that causes injury it is more than likely that any action for civil damages brought by the injured party or his dependants against a contractor or designer would be brought under the tort of negligence.

Tort
The rule of common law is that everyone owes a duty to everyone else to take reasonable care not to cause them foreseeable injury. Hence a contractor will be liable if he fails to take reasonable care to protect those on site from foreseeable injury, and a consulting engineer will be liable if he fails to take reasonable care to prevent injury by reason of a faulty design. The basis of this liability is the tort of negligence, and the remedy or redress is an assessment by the courts of the damages suffered.

Negligence
Negligence within the context of health and safety is an independent tort consisting of a breach of a legal duty to take care resulting in damage. In the famous case of *Donoghue* vs. *Stevenson* (1932) A.C. 562, Lord Atkin said, 'You must take reasonable care to avoid acts or omissions which you can reasonably foresee would be likely to injure your neighbour. Who then, in law, is my neighbour? The answer seems to be — persons who are so closely and directly affected by my act that I ought reasonably to have them in contemplation as being so affected when I am directing my mind to the acts or omissions which are called in question'.

Limitation in tort
Whereas under contract law the period for which a contractor or engineer remains liable is six years in the case of an action on a simple contract and 12 years for a contract under seal, in tort the usual period of limitation is six years.

The cause of action in the case of tort accrues at the moment when the plaintiff would be entitled to succeed in an action against the defendant. In the case, for example, of negligence, this moment will be as soon as the damage occurs. When damage has arisen the cause of action accrues even though the plaintiff may be unaware that the damage has in fact occurred.

In the case of personal injuries, legal proceedings must be started within three years of the accident. After that time they are statute barred. This time may be extended in special cases, for example, to those who are found to be suffering from certain diseases after the limitation of time has expired.

The period of limitation has been modified to some extent by the Latent Damage Act 1986. The main effect of this Act is that, in negligence cases involving latent defects, the existing six-year period of limitation is subject to an extension to allow the plaintiff three years from the date of discovery or reasonable discovery of significant damage. However, a long stop is applicable, which bars a plaintiff from initiating court action more than 15 years from the defendant's breach of duty.

Negligence involving injury

The incident at Abbeystead near Lancaster when an explosion in a completed pumphouse killed 16 people and injured many others illustrates the process of the law of tort in civil engineering. Methane had entered the valve house from an underground void.

The Water Authority, consulting engineer and contractor were sued in the High Court for negligence by a consortium of 31 victims and relatives. The contractual relationships between the defendants were irrelevant as they were sued in tort and the plaintiffs claims turned on the common law duty of care.

Following an appeal by a majority of two to one, the judges found the consulting engineer solely to blame for the accident. The Appeal Court ordered the consultant to pay the costs of all parties for both the Appeal and the previous High Court rulings.

Employer and employee

The relationship between employer and employee is covered in a form of contract known as an employment contract but liability for injuries, disease or death at work is normally dealt with under tort. In order to show that an employer was negligent, an employee must establish that the employer

- owed him a duty of care
- acted in breach of that duty
- caused injury, disease or death to the employee.

The duties of employers at common law were identified by the House of Lords in the leading case of *Wilson's and Clyde Coal Co.* v. *English* (1938) A.C. 57. All employers must provide a competent workforce, adequate materials and a proper system and effective supervision. The employer's duty is not absolute; he does not warrant that his premises and the conditions of work are safe. He is bound only to take all reasonable care, but if he delegates the performance of his duty to an independent contractor he still remains liable for the negligence of that contractor.

The common law duties of employers are owed only to employees. They are not owed to employees of independent contractors (as distinct from subcontractors) who may be working on the same site, even though breach of duty towards persons other than employees gives rise to criminal liability under the HSW Act (1974). The fact that a person styles himself as self-employed does not mean that he cannot be an employee, and it follows that a contractor may be held liable for injuries suffered by a labour-only subcontractor in spite of the fact that the latter regarded himself as self-employed.

Disobeying safety orders

An employee who disobeys safety instructions can be dismissed from his job after due warnings, or instantly in the case of gross misconduct. To ensure that dismissals on these grounds are reasonable, it is prudent for every employer to include in employees' terms of engagement, or in the health and safety policy statement, a warning to the effect that disobeying safety instructions and failing to co-operate with their employer in fulfilling his statutory duty on health and safety are subject to reprimand and, in repeated or extreme cases, dismissal.

In the case of employment injuries, the defence by employers of *volenti non fit injuria* (to one who is willing no harm be done) is rarely successful, the courts taking the view that the dangers of employment are not voluntarily accepted by workers; rather that employment is an economic necessity. However, an employer may be successful in reducing damages on the grounds of contributory negligence by the employee.

Vicarious liability

Vicarious liability is the principle under which a person is held liable for the tort of another committed against a third party. Under this doctrine an employer is liable for the torts of his employees committed by them in the course of their employment whether authorised by their employer or not. If an employee, while acting in the course of his employment, negligently injures another employee or an employee of another company working on the site, or even a member of the public, the employer will be liable for the injury. Vicarious liability is the ground on which most claims for injury-causing accidents are successful.

3

Enforcement of health and safety law

Application to Great Britain
Before 1 January 1975 when the Health and Safety Commission and Executive were established, HM Factory Inspectorate within the then Department of Employment was responsible for implementing safety, health and welfare law applicable to the construction industry. Although since that date the Factory Inspectorate still has that responsibility, it now operates within the framework of the Health and Safety Executive. There are, in addition, some aspects of safety in civil engineering (e.g. mines and quarries, nuclear installations, and hazardous substances) which are covered by other inspectorates, and others (e.g. railways) that are administered by the Commission on behalf of other ministries.

Health and Safety Commission
The HSW Act makes provision for the establishment of a Health and Safety Commission. This is the authority in England, Scotland, Wales and offshore that is responsible to the Secretary of State for Employment for all health and safety at work matters. Its duties include promoting the objectives of the HSW Act, carrying out and encouraging research and training, providing an information and advisory service and putting forward to Government proposals for regulations under the Act. The Health and Safety Agency for Northern Ireland has similar duties and powers in Northern Ireland.

The Act requires the Commission to consist of a full-time Chairman appointed by the Secretary of State for Employment, and between six and nine part-time members. Three of the part-time members are required to represent employers' organisations, three employees' organisations and three local authorities and other appropriate organisations.

Health and Safety Executive
The HSW Act also makes provision for the establishment of the Health and Safety Executive (HSE). This body consists of three full-time members, one of whom is the Director General. They are appointed by the Health and Safety Commission with the approval of the Secretary of State.

One of their duties is to make adequate arrangements for the enforcement of the relevant statutory provisions. Thus the enforcement of health and safety at work law is undertaken by the staff appointed by the HSE.

ENFORCEMENT OF HEALTH AND SAFETY LAW

The other bodies which assist the Commission in a similar way include the local authorities, which have a special responsibility for safety and health in offices, shops and warehouses; the Railways Inspectorate of the Department of Transport in respect of railway employees (*see* Fig. 3.1) and the Petroleum Engineering Division of the Department of Energy in respect of most aspects of safety in connection with offshore oil and gas.

Figures 3.1 and 3.2 show the reporting relationships of the Health and Safety Commission, the Executive and its various Inspectorates, and regional organisation. Appendix 3 gives the HSE addresses.

HM Factory Inspectorate. The Factory Inspectorate of HSE is the enforcement authority responsible for the statutory requirements for safety, health and welfare on construction sites. It also has other enforcement duties, for example in respect of factory premises, hospitals, schools, universities and fairgrounds. It is supported and assisted in this by all the other divisions of HSE, but particularly by the Technology Division, the Research and Laboratory Services Division and the Medical Division.

For field inspection purposes of the Factory Inspectorate, of Britain is subdivided into twenty geographical areas, each under the control of an area director (*see* Fig. 3.2 and p. 32). Areas are staffed by inspectors who are involved in the day-to-day inspection and advisory activities. In each area one group of inspectors, under the direction of a senior inspector, is solely concerned with the activities of the construction industry. They carry out unannounced inspections of sites and investigations of accidents or dangerous occurrences on site, and provide advice on compliance with the appropriate law to contractors and others who have specific interests in construction activities, for example manufacturers of construction plant.

In addition, the National Industry Group for Construction is located in the London south area office. This co-ordinates the activities of the construction groups in the areas and also acts as the focal point for discussion between the construction industry and HSE.

The Technology Division. The Technology Division of HSE, which is staffed by professionally qualified engineers, chemists, occupational hygienists, physicists and microbiologists, also has professional responsibility for a field force in addition to its larger headquarters units. For this purpose Britain is divided into seven geographical areas each covered by a Field Consultant Group (FCG)(*see* Fig. 3.2). Conventionally each FCG (except the one covering the north-east of England) is housed within one of the Factory Inspectorate area offices and serves two or three adjacent areas. In addition to mirroring the headquarters disciplineunits, they have a nucleus of scientists and laboratory facilities for carrying out analysis and sampling work (field scientific support units). They provide immediate professional and technical support to the field inspectors.

Staff in headquarters units of Technology Division specialise in various topics within their discipline, e.g., in the Construction Unit, tunnelling, structural engineering, demolition work including the use of explosives, falsework, scaffolding and construction plant. They provide support and authoritative professional advice to HSE's field and policy divisions in

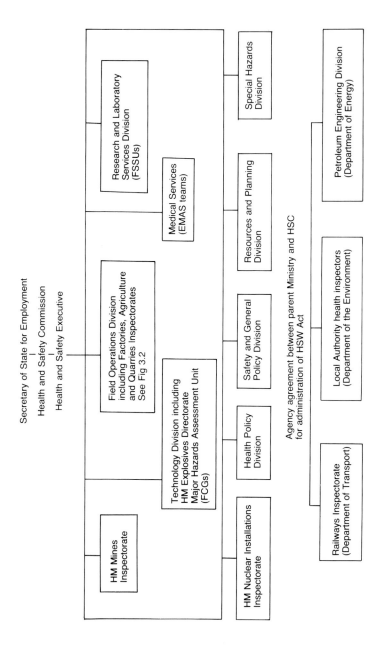

Fig. 3.1. Reporting relationships within the HSE (The Railways Inspectorate joined the HSE on 6 October 1990)

ENFORCEMENT OF HEALTH AND SAFETY LAW   31

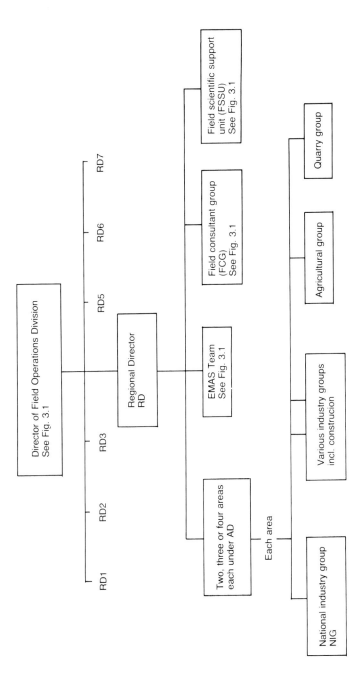

Fig. 3.2. *Regional structure of the HSE*

addition to contributing nationally and internationally in safety and health matters as members of British and International Standards Committees.

The Technology Division includes HM Explosives Inspectorate, which is concerned with the manufacture, transport, handling and security of explosives, and the Major Hazards Assessment Unit. The unit develops techniques and methods for assessing quantitatively hazards and risks arising from major hazards. It formulates advice for use by planning authorities for the siting of new major hazard premises and for the control of developments near existing major hazard sites. Major hazard sites may briefly be described as plant or premises which contain sufficient quantities of hazardous substances with the potential to cause major accidents from the release of flammable or toxic substances (*see also* p. 195).

HM Nuclear Installations Inspectorate. HM Nuclear Installations Inspectorate licenses nuclear installations ranging from nuclear power stations and nuclear chemical works to research reactors.

The Medical Division. The Medical Division of HSE is staffed by doctors and nurses who specialise in occupational medicine. They have both a headquarters and a field presence. In the field they are based on area offices and provide advice on health at work; they also carry out surveys and investigations on occupational ill health. The Employment Medical Advisory Service (EMAS) provides free advice at the request of any employed or self-employed worker, trade union representative, employer or medical practitioner on the effects of work on health, together with guidance on the placement and return to work of people with health problems.

Field Operations Division. From April 1990 a regionalised structure came into effect for the management of HSE field operations. This resulted from a review carried out early in 1989. Seven regional directors of field operations (RDs) are each responsible for a number (between two and four) of the 20 HSE areas each under its area director (AD). The factory, agricultural and quarries inspectors form part of the AD's team.

Each region also includes EMAS teams (one for each area), one field consultant group (FCG) and one field scientific support unit (FSSU). Each RD reports to the Director of Field Operations, who in turn reports to the deputy Directory General of HSE. The field management structure is shown in Fig. 3.2.

Powers of the Factory Inspectorate

The task of the inspectors within the construction groups is to ensure compliance with the HSW Act by all those engaged in construction operations within their particular Factory Inspectorate area. They have wide powers to enter sites and premises to carry out inspections, examinations and investigations. These powers include the right to

- enter premises or construction sites at any reasonable time to conduct inspections and examinations for the purpose of ensuring compliance with any of the relevant statutory provisions

- take with them any other person who has been authorised by the Executive (e.g. another inspector, a scientist or a photographer), and any equipment or materials required to carry out their work
- carry out investigations and examinations (e.g. in respect of an accident)
- direct that any premises or site (or any part of them) and/or anything in them such as plant, machinery, substances, etc. be left undisturbed until their investigations have been completed
- take measurements, photographs and any recordings they consider necessary for the purpose of any examination or investigation they undertake
- take samples of articles, substances and the atmosphere in or in the vicinity of the premises and site they are empowered to enter
- require any article or substance to be dismantled or subjected to any process or test if they consider it to have caused or be likely to cause danger to health or safety; however, this should be carried out without damaging or destroying the article or substance unless necessary for the purpose of carrying into effect any statutory provision
- take possession of any article or substance mentioned in the previous item above and keep it for as long as is necessary in order to
    - examine or dismantle it or both
    - ensure it is not tampered with before their examination has been completed
    - ensure it is available for use as evidence in legal proceedings in which event the inspector is required to leave with a responsible person on the premises, a notice describing the article or substance, sufficient to identify it and stating that the inspector has taken it; if, in the case of a substance, it is possible, he shall only take a sample, marked in a manner sufficient to identify it
- seek information from any person whom they consider may be able to help with any examination or investigation they are conducting, require the person to answer questions (in respect of the investigation etc.) and require the person to sign a declaration of the truth of the answers; this may only be carried out in the presence of a person nominated by the person being questioned and any person the inspector may allow to be present; however, the answers given to the inspector cannot be used in evidence in any proceedings against that person for a breach of the Act or relevant regulations; also they cannot be used in evidence for similar proceedings against the husband or wife of a person
- require the production of, inspect and take copies of or copy any entry in
    - books, documents, registers, etc. that are required to be kept by the relevant statutory instruments
    - any other books, documents, drawings, etc. which it is necessary for them to see for the purposes of any examination or investigation they are undertaking

- require any person to provide such facilities and assistance with matters or things within that person's control or in relation to which that person has responsibilities as are necessary to enable the inspectors to exercise the powers conferred on them
- exercise any other powers necessary for the purposes of their work
- seize and cause to be rendered harmless, by destruction if necessary, any article or substance in the premises or on the site that they consider, in the circumstances seen, could be a cause of imminent serious personal injury.

Methods to ensure compliance

Various ways are used by the inspectors in the field to secure compliance with legal requirements. Many involve routine inspection of sites to check what is occurring against what is required by the law. The visits are made without giving any prior warning to the contractor or site manager.

Advice and persuasion. The inspector may find a contravention of, for example, the Construction Regulations. In some cases the contractor may not be aware of the particular requirement or how it may be complied with. The inspector will then provide advice on ways in which the obligation may be met. However, the contravention may be an oversight by the contractor or something unknown to the person in charge. Again, advice and persuasion to correct the fault will be the inspector's first action.

If the inspector finds a number of matters requiring his attention, again verbal advice will be given on what is wrong and how to put it right. This may then be put in writing to the firm concerned, indicating also the various breaches of the law.

Notices. The inspector is provided with a particularly quick and effective means of dealing with breaches of the law and hazards that may be found during site inspections. He can issue an official notice. There are two types: the Improvement Notice and the Prohibition Notice.

An *Improvement Notice* can only be issued where an inspector considers that there has been a contravention of a statutory provision and it is likely that the contravention will continue or be repeated, or at the time of his visit a contravention (or more than one) is seen. An example of an Improvement Notice is given in Fig. 3.3. The Inspector has to include in the notice

- the location of the site or place where the contravention occurred
- the particular provisions contravened
- the reasons why in his opinion the provisions have been contravened
- a realistic date by which the matters are expected to be remedied.

The contractor may appeal against the notice but this must be done within 21 days of the date of issue. The appeal has to be forwarded to an Industrial Tribunal (the address of the Secretary of the Tribunals, together with the procedures for lodging an appeal, are given on the Notice) which after consideration may cancel, confirm or modify the notice. When an appeal is lodged the requirements of the notice are suspended pending the results of the appeal.

A *Prohibition Notice* may be used only when the Inspector considers that there is a risk of serious personal injury arising from an activity or operation being carried on or about to be carried on. They have the effect of preventing an activity or operation being carried out or continued until the requirements specified by the inspector in the notice have been carried out and any associated contraventions remedied. An example of a Prohibition Notice is given in Fig. 3.4.

The notices can be issued to have an immediate or a deferred effect. The immediate type may only be issued if the inspector considers that the risk of injury is imminent. If it is of the deferred type then the prevention of the activity or operation becomes effective from the date stipulated in the notice unless the requirements of the Inspector have been complied with. With both types of notice, when the contractor has remedied the matters and any associated contraventions that are stipulated, the notice ceases to have any effect. As with Improvement Notices, an appeal may be lodged through an Industrial Tribunal, but Prohibition Notices remain in force until the appeal has been heard.

Prohibition Notices are served in the anticipation of danger and may be issued without there being a contravention of any particular regulation. However, if there has been a legal contravention then the details must be stipulated in the notice.

Both the immediate and deferred notices may, but need not, include directions for remedying the faults or contraventions which the inspector requires to be carried out.

Prosecution. The ultimate course taken by inspectors to uphold the law is to prosecute for contravention. Anyone who does not comply with any part of the HSW Act or its relevant regulations can be prosecuted in the Magistrates' Court (or in the Sheriff's Court in Scotland). The maximum fine which may be imposed for each offence if the defendant is found guilty is £2000. However, serious cases may be brought on indictment before the Crown Court. In such cases if the defendant is found guilty a prison sentence of up to two years or a fine (the amount unlimited) or both may be imposed.

An example of an offence that could invite trial in the higher court is that of failing to comply with a Prohibition Notice.

Consultation. Area inspectors also act in a consultancy capacity. They visit, by invitation or otherwise, contractors' offices to discuss problems, and to advise on legal requirements and on how best to meet these obligations. They attend local safety group meetings and give talks and advice to members on health and safety matters. They also call together representatives of specialist contractors (e.g. roofers, demolition contractors and scaffolders) in their areas to discuss ways of meeting their legal obligations.

Other activities

In addition to the work of inspectors in the field, there is continuous activity by inspectors and others from the various divisions at headquarters of HSE to help contractors and others concerned with the

36 CONSTRUCTION SAFETY HANDBOOK

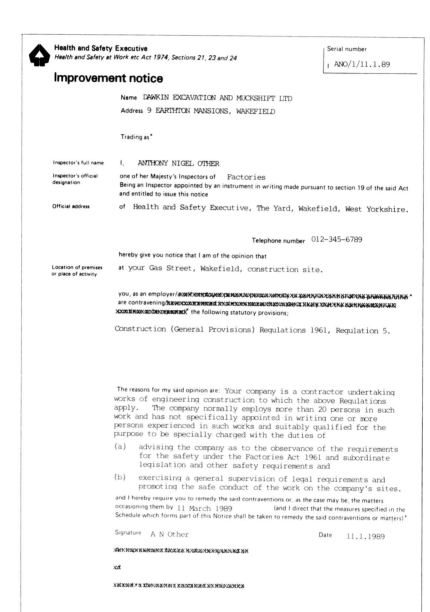

Fig. 3.3 (above and facing page). Example of an HSE Improvement Notice

HEALTH AND SAFETY EXECUTIVE
Health and Safety at Work etc. Act 1974, Sections 21, 22, 23, and 24

Schedule

Serial No. I/P

If any of the following measures are put into effect then the Notice will be deemed to have been complied with.

Either:

1. Any person or persons in the company with suitable experience of site work should be sent on the "Construction Safety" Course at XYZ Polytechnic or the "Safety Supervisor" Course of the South Riding Safety Group and then appointed in writing as a company safety supervisor with responsibility to visit regularly and inspect all company construction sites

Or:

2. At outside specialist safety management company shall be appointed by Dawkin Excavation and Muckshift Ltd to undertake on Dawkin's behalf the duties of Regulation 5.

Or:

3. Other action shall be taken by the company of equivalent standard to 1 or 2 above to ensure compliance with Regulation 5.

A N Other   11.1.1989

---

construction industry to reduce accidents and ill health on site. This includes
- the collection, analysis and publication of accident statistics
- the production of guidance notes and other information (*see* p. 200 and Appendix 4)
- liaison with the various national trade and industry federations and trade unions
- representation on British Standard committees, on the Safety in Civil Engineering Committee of the Institution of Civil Engineers, the International Tunnelling Association Committee and the British Tunnelling Society Committee, for example
- liaison with other professional institutions.

CONIAC. A particularly important activity is the involvement with the Construction Industry Advisory Committee. The Committee consists of representatives from both the employer and the employee sides of industry, and its function is to advise the Health and Safety Commission about the protection of people in the industry and the public from hazards

38 CONSTRUCTION SAFETY HANDBOOK

Fig. 3.4. Example of an HSE Prohibition Notice

to health and safety arising from construction activities. It is chaired by a senior factory inspector and attended by another who is head of the National Industry Group for construction, the specialist inspector who is head of the construction unit within the Technology Division, and a senior doctor from the Employment Medical Advisory Service. CONIAC has been responsible for a number of useful publications, some of which are listed in Appendix 4.

4

Hazards of construction and their prevention

From the analysis of injuries and causes of ill health described in chapter 1 it will be evident that providing safe working places and safe systems of work by effective safety management is the single most important contribution engineers and other professionals can make to reduce the number of deaths and injuries in the construction industry. In this chapter the common causes of accidents are examined and methods suggested for their prevention.

Part 1 is concerned with the risk of physical injury and Part 2 with the risk of ill health.

PART 1. PHYSICAL INJURY
Excavations

For convenience this section examines the hazards and protective measures concerned with three types of excavation work: trenches; basements and wide excavations; and shafts.

Trenches

The main cause of accidents in trench work is the collapse of the sides burying or partially burying those in the trench (Fig. 4.1). About 1 in 5 of this type of accident results in death. The other major causes are materials falling into the trench striking those below and also both plant and people falling into the excavation.

Some of the reasons why the collapse of a trench is so dangerous are as follows.

- The trench provides a very confined working space, and falling materials cannot easily be avoided.
- Collapse occurs quickly and usually without warning.
- The likelihood of locating, uncovering and rescuing any buried worker in time is not good. A hydraulic excavator should never be used to clear away the debris for fear of causing further serious injury to the buried person. Even hand tools such as spades have to be used with great care to avoid further injury.
- The rescuers themselves have to be safeguarded against further collapse and the provision and positioning of extra shoring for this takes valuable time.

HAZARDS OF CONSTRUCTION AND THEIR PREVENTION

Causes of trench collapse. The most common causes of trench collapse are

- mechanical failure of the soil
- breakdown in strength of the ground by the ingress of water
- vibration and weight of vehicles and plant
- storage of materials and equipment near the edge of the trench
- variation in the nature of the ground
- previously excavated ground
- the sides of the trench being struck by heavy loads.

When a trench is cut, lateral support on the cut planes is removed and *mechanical failure* occurs if the ground is unable to support its own weight. Similarly, slipping and collapse may occur in rock excavations when bedding planes dip steeply across the line of the excavation. The use of explosives in the excavation process may well trigger off a slipping movement.

Fig. 4.1. Typical trench collapse due to inadequate shoring and excessive surcharge, often causing fatal injuries (courtesy HSE)

The *ingress of water* not only removes particle adhesion or lubricates slipping planes but also increases the weight (pressure) on the unsupported walls of the trench. Water penetration into cracks because of shrinkage of clays in hot dry weather causes softening and a rapid loss of strength. Another effect of water percolation, during winter months, is that the water may freeze and cause a breakdown of strength, thus increasing the size of any cracks or fissures.

The *movement or operation of vehicles and plant* (e.g. dumpers and cranes) close to the edge of a trench produces a heavy surcharge load on the unsupported sides which they are unable to resist. Vibration due to movement of plant may cause particle movement or even displacement of stones and small boulders which in turn leads to instability. For trenches in or adjacent to roadways, normal traffic is a particular hazard.

Excavated spoil, pipes, bricks and timbering equipment *stored too close to the edge of a trench* may produce a surcharge load which is too great for the trench sides to withstand. No materials or spoil should be placed within a metre of the trench and the height of spoil heaps near the trench should be limited to between 1 and 2 m.

Excavation is rarely in homogeneous ground. *Differing strata* may occur in relatively small vertical distances and inclusions of one type of soil may occur in another, e.g. pockets of sand may occur in a predominantly clay soil. Various sizes of boulder may occur, e.g. as in boulder clay, and sometimes there are pockets of water. Such mixtures considerably reduce the strength of the ground and limit its ability to stand unsupported for any length of time.

Previously excavated ground reduces the inherent strength of the adjacent ground — the backfill may have been of variable consistency and not too well consolidated. When previously excavated ground is close to or crosses the new trench collapse may well occur. Examples of this are as follows.

- When pipes cross a newly excavated trench the ground in the vicinity of a previous excavation will be very weak and particularly susceptible to failure.
- If a previously excavated trench exists approximately parallel to and within a meter of the new trench, the ground between them will be weak and very likely to collapse.
- A further example, not exactly of a previous excavation, is that where hedges, bushes or trees are in close proximity to the line of the new trench whether or not they have been grubbed up before excavation starts. The action of the roots in taking moisture from the surrounding ground causes a breakdown in the strength of the soil.

The *sides of the trench may be struck by heavy loads* during the lowering of pipes or other materials into the trench.

When a trench is excavated close to and to a depth equal to or below the foundations of a building or structure the *surcharge load* may not only cause the collapse of the sides of the trench but also may well cause the collapse or partial collapse of the structure.

A further common cause of collapse is *insufficient and poor quality timbering* used to support the trench sides.

Prevention of trench collapse. The likelihood or otherwise of a trench collapsing can rarely be assessed by visual inspection. There is almost no ground that will not collapse given certain conditions (as illustrated above) and visual inspection cannot ensure that none of these will arise.

Sometimes, following a visual inspection, an assumption is made that the trench sides may well be self supporting for a short period. This is a fallacy and should never be risked. Many fatal and serious accidents have occurred in untimbered trenches which have only been open for an hour or two — a substantial proportion have occurred during the day of the excavation. In 1971 Tomlinson[1] referred to experiments he had made relating to the stability of the vertical sides of a trench on a site where the soil consisted of soft to firm alluvial clay 1·5–2 m thick overlying soft to very soft peaty alluvial clay. The results showed that for various depths from 2·9 to 5·5 m, partial collapse occurred after 14–37 min and complete collapse after 23–49 min. The longest times, 37 min and 49 min were for the shallowest excavation.

The ways to prevent collapse and make working in excavations safe are either by

- timbering or shoring, i.e. supporting the sides (the provision of physical support for each side of the trench)
- battering the sides to a safe angle.

Battering requires the sides of the excavation to be sloped no more than the angle of repose for the soil. This would obviously vary for the particular ground but in no cases should the angle be more than 45° to the horizontal and often much less, e.g. nearer 30°. A guide to the safe angle at which homogeneous soils will stand is given in *Timber in excavation*.[2]

Sometimes a combination of timbering and battering is used, usually in fairly deep trenches: for the first metre or so from the bottom of the trench, the sides are vertical and then they are battered to a safe angle. In this case the vertical sides must be supported as for timbering.

Supporting the sides of a trench may be carried out with timbering, H-piles, drag boxes or proprietary systems.

Conventional *timbering* comprises open poling or close sheeting supported by walings strutted across the trench — depending on the type and quality of the ground. Although this method is called timbering, sometimes steel trench sheets are used for poling boards and adjustable props rather than timber are used for struts.

H-piles are steel universal column sections pre-driven into the ground at intervals along the intended line of each side of the trench with their flanges parallel with each line. As excavation proceeds the exposed sides are supported by timbering or trench sheets laid horizontally behind the flanges of adjacent piles and wedged in position against the ground.

Instead of conventional timbering *drag boxes* may be used, which usually comprise steel framed boxes with strong side sheeting and struts. The two ends, top and bottom are open and the width is slightly less than

Fig. 4.2. Proprietary prefabricated trench support system (courtesy Shephard Hill and Co. Ltd)

that of the trench. The two sides extend to the full depth of the trench and are set on skids. They have to be strong enough to withstand any collapse of the trench sides. Men may then work at the bottom of the trench within the boxes in safety. As trench bottom work proceeds, the boxes are pulled along, usually by the excavator, and the men remain inside within the protection of the side sheeting.

Various *proprietary systems* such as hydraulic struts and associated vertical poling members are in use for trenching (Fig. 4.2). Similarly, a hydraulic strut and waling frame system is used, in conjunction with ordinary trench sheets used as poling boards. A third type consists of specially formed vertical members with slots and incorporating struts. These are installed in facing pairs at intervals along the trench and heavy side plates are slid into the slots to the full depth of the trench.

Before timbering is carried out or proprietary systems are used all material and equipment should be inspected to ensure that it is sound and suitable. With the installation of traditional timbering no work should be done outside the protection of the already constructed support. All installation work must be carried out from above or behind the completed work or from suitably strong prefabricated boxes which may be lowered into the trench and from which workmen can work in safety. Many fatal accidents have occurred to men installing the timbering when working no more than 0·5 m outside the previous support. The great advantage of the proprietary systems is that they can be installed from ground level without the need to enter the danger area of the trench.

There is a need to inspect and maintain the timbering because climatic conditions and work operations may affect its strength and stability, e.g. wedges may work loose. In Britain the Construction (General Provision) Regulations 1961 require that with an excavation over 1·2 m deep every part of it must be inspected by a competent person at least once every working day. Also a thorough examination of the trench and all the support work, timbering, etc. by a competent person must be made every seven days and the results recorded in a register. The working end of trenches more than 2 m deep must also be inspected at the start of each shift.

Other hazards of trench work. Other hazards associated with trench work not involving soil collapse include

- workers being struck by falling materials
- workers being struck by falling rocks
- people falling into the trench
- unsafe means of access
- workers being struck by excavating machinery
- vehicles falling into the trench.

From time to time workers in trenches are *injured by falling materials* which have been lying about at ground level and are accidentally knocked into the trench or slip into it, moved perhaps by the vibration of passing vehicles. The materials may be pipes, bricks, adjustable props, trench sheets or timber, etc. intended for use in the trench. The accidents may be very severe (e.g. in a trench 2 m deep a pipe falling on to a man bending down may fracture his spine). Materials should not be left anywhere near the trench but should be properly stacked well clear. They should be taken to the trench only when ready for immediate use and then passed to those in the excavation.

Workers may be *struck by rocks* falling from the sides of the trench; the rocks may be loosened first during the excavation process and subsequently by vibration or from being struck when materials are lowered into the trench. This often applies when trenches are cut in existing roadways and the boulders in the sub-base are loosened or where rock blasting has been carried out. During timbering and before work in the trench begins, it is important that all loose rocks, boulders and stones, etc. should be identified and secured or removed. Also, that workers in the trench should wear safety helmets at all times.

Where people are likely to suffer *falls* of more than 2 m into a trench rigid barriers between 915 mm and 1·05 m high must be provided. This may be done by extending the timbering above ground level and fixing suitable guard rails and toe boards to it. Alternatively, a properly constructed barrier fence parallel to and a short distance away from each edge of the trench should be erected. A post and guard rail type would be sufficient. In across-fields work where few people are likely to be in the vicinity a spoil heap with rope or brightly coloured bunting carried by posts at intervals on top of the spoil would be acceptable. However, even if the depth of the trench is less than 2 m it is advisable to take similar precautions to prevent anyone falling into it. Lighting or warning lights should be provided at night where appropriate. This is particularly important where the trenches are in public places.

Access may be required across the top of a trench at ground level. The access should be properly constructed and should be at least 430 mm wide for pedestrian use only — where it is intended to carry materials across (e.g. in a wheelbarrow) the width must be not less than 600 mm. Where the distance from the walkway to the trench bottom is greater than 2 m each side of the walkway must be provided with guard rails and toe boards.

Ladder access should be provided at reasonable intervals within the part of the trench where the sides have been adequately supported. On no account should a ladder that could be used be placed in an unsupported area of a trench. Ladders for access should be sound and properly constructed, should project at least 1·05 m above ground level and be tied to prevent slipping.

To prevent workers being *struck by excavating machinery* a safe method of working needs to be implemented. During the excavation cycle no person should be allowed in the vicinity of the machine. People in the trench should be well away from the face and those at ground level kept outside the slewing radius. When the excavator driver cannot see all parts of the jib and bucket during the excavation cycle or when the machine is used as a crane to lower materials (e.g. pipes) into the trench, an experienced banksman should be used to guide the driver and to ensure that people remain well clear of the operations being undertaken.

In the most common occurrence where *vehicles fall into a trench* small site dumpers — the front-end tipping type or other vehicles used for tipping materials (e.g. concrete) into the trench — either fail to stop at the edge or the ground gives way as they tip their load into the trench. The vehicles should be prevented from approaching too close to the edge by the use of adequate stops. These may be securely anchored timber baulks where the anchorage arrangement is made at a distance of twice the depth of the trench from the baulk near the edge. Proprietary stop block devices are also available (*see* Fig. 4.27). With this operation a banksman should always be employed to guide the driver.

Other reasons for this type of accident may simply be the inattention of the driver or too fast an approach. This points to the need for tighter supervision. It is essential that the driver is competent and should there be any doubt about his ability he should be given suitable training before being allowed to continue with the type of work in question.

Brake failure as a possible cause of accidents may be due to inadequate maintenance and this must be corrected. Sometimes a site vehicle such as a crane travelling about the site may approach too close to the edge of a trench, causing the ground to collapse. Where heavy vehicles are likely to travel close to trenches a barrier must be erected to prevent such approaches being possible.

Another type of vehicle-in-trench accident occurs in trenches alongside or in a working carriageway. With this type of trench it is essential to erect suitable warning signs well before, and at the approach to, the excavation. Also, suitable barriers and coning should be erected at the approaches.

HAZARDS OF CONSTRUCTION AND THEIR PREVENTION 47

Fig. 4.3. Use of contiguous piles as permanent ground support before excavation at Channel Tunnel portal through Castle Hill, Folkestone (courtesy TML and Eurotunnel)

Basements and wide excavations

Most of the hazards of trench work are also applicable to basements and wide excavations. However, supporting the sides of a deep and wide excavation may be safely carried out by the following methods.

Before excavation a diaphragm wall is sometimes formed around the perimeter of the site. This is designed to act as a retaining wall for the sides of the excavated basement. Following the construction of the wall excavation may then take place to the necessary reduced levels in perfect safety for the operatives, since the sides are prevented from collapsing. A similar system also in use is that of contiguous bored piles (Fig. 4.3) where a continuous line of bored piles to a depth below formation level is installed around the perimeter of the site. These are permanent supports and their use may have been decided at the design stage (*see also* p. 166).

Another method is by means of close boarding with timber, trench sheets or even sheet piles (Fig. 4.4). With timber runners or trench sheets these are installed progressively in a narrow excavation around the edges of the site. The runners are used in conjunction with walings at suitable vertical centres and the ground pressures transferred via the walings through a system of raking shores and thrust blocks at their lower ends,

Fig. 4.4. Proprietary hydraulic walings used to brace a 6 m deep-thrust bore pit (courtesy Mechplant Ltd)

to the unexcavated ground. When the close boarding has been installed the rest of the excavation can be carried out in safety.

When constructing basements, deep excavations and shafts, it is particularly important to construct a solid fence around the site before excavation is started.

Shafts

With shaft construction the hazards are similar to those for trenches, i.e. materials falling into the shaft from ground level, earth or boulders falling from the sides and people falling into the excavation. Support of the sides during shaft construction may be carried out with traditional timbering installed progressively with the shaft excavation. Where the timbering is properly carried out there is no danger to the workers in the shaft either from the collapse of the sides or from dislodged boulders. At the top, the timbering should be extended to form a fence so that people and materials are prevented from falling into the excavation. Where the timbers are not extended in this manner a close-boarded fence should be erected around the shaft top.

Shafts of larger diameter may be constructed with precast concrete lining rings installed as excavation proceeds, thus providing constant support to the sides. Where excavation is to proceed at a faster rate than lining installation some form of temporary support for the sides must be used. One method is to use sprayed concrete. This may introduce a health hazard from the overspray and rebound, so the operators must use full protective clothing and appropriate respiratory protective equipment.

Workers should only be conveyed down a shaft to their place of work by means of a man-riding hoist or suitable skip. Skips should be at least 1 m deep and provided with means to prevent occupants falling out and also with means to prevent the skip spinning or tipping. They should only be lowered under power by a suitable crane or winch.

Effective communications. With deep shafts the maintenance of an efficient communication system between the people working in the shaft and those at the top is important to ensure the safe control of the lowering and lifting of skips or other conveyed loads. Control is usually best vested in an experienced banksman at ground level who should ensure during lifting and lowering that people at the lower level in the shaft keep well away from the area directly below the load. Where a crane is used the crane driver must also be in the communication net and receive the necessary guidance from the banksman.

Dangerous atmospheres. Another type of hazard associated with shaft construction is that of dangerous atmospheres, which may result from

- oxygen deficiency
- carbon dioxide
- carbon monoxide
- nitrous fumes
- methane gas.

Tests for *oxygen deficiency* should be made before anyone enters the shaft at the beginning of each day. There should also be continuous testing while work is being carried out. In some types of ground a fall in the barometric pressure may cause a sudden change in the atmosphere to one severely deficient in oxygen.

Although an excess of *carbon dioxide* may be found in any ground it is particularly likely in chalk and limestone areas. Again the atmosphere should be tested at regular intervals for the presence of carbon dioxide.

Where internal combustion engines are used in the excavation the exhaust fumes will contain *carbon monoxide*. In these cases not only should the atmosphere be continuously tested but also an efficient ventilation system must be introduced in the shaft to provide a safe working atmosphere.

Nitrous fumes would occur if the use of explosives were resorted to during the construction process (e.g. if a band of very hard rock were encountered). Obviously during the firing and immediately afterwards no one will be in the shaft; nevertheless an adequate ventilation system must be provided to clear the nitrous fumes and to ensure a safe working atmosphere for the returning workmen. The atmosphere must be tested before work starts to ensure that it is fit to breathe.

Methane gas may be encountered in or near coal measures and carbonaceous ground, and in or near refuse dumps. It is flammable and highly explosive when mixed with air at concentrations of 5–15% by volume (of methane). Constant monitoring of the atmosphere is essential when there is any possibility that methane may seep into the excavation. Where concentrations cannot be kept below 0·25% by volume only explosion-protected electrical equipment together with an adequate system of ventilation must be used. However, if the concentration reaches 1·25% by volume all men should be withdrawn from the workings.

Deep bored piles

With deep bored piles there are similarities to that of shaft construction in respect of the safety and health of the workforce. For inspection purposes, for bottom preparation such as cleaning out and sometimes for hand-digging extensions to under-reams, workmen are required to enter deep bored pile excavations. However, entry must never be made into boreholes with a diameter of less than 750 mm.

Those who descend boreholes and those who work in shafts or trenches encounter similar dangers. Materials or tools may fall down the shaft from ground level or men may fall into the borehole from the surface. The sides may partially or even completely collapse and there is a likelihood that atmospheres are unfit for breathing. Where work has to be carried out in an under-reamed section there is the possibility of the roof collapsing.

Prevention of items from falling. All boreholes left unattended should be securely covered or have an adequate fence, reasonably close to the edge at ground level, to prevent people falling into them. No tools, materials or loose spoil should be allowed within a metre of the borehole at ground level and an upstand of at least 150 mm should be provided close to the edge. This may be carried out by allowing the lining tube to protrude above ground level. Another precaution to help ensure that tools do not fall down the shaft is for those people who descend the borehole or who work near it at the surface to be banned from carrying loose tools or equipment.

Entering a large diameter pile. Descent should only be made in specially designed safety cages for inspection purposes or in open muck skips fitted with an anti-spin device, and all lowering and lifting must be carried out under power control by means of a suitable crane or winch. While anyone is below ground the lifting appliance must be in the immediate charge of a competent operator and the power source kept running. Unless the operator of the lifting appliance can see the people working at the bottom of the borehole at all times a competent banksman must be employed at the top of the borehole to take charge of all lifting and lowering operations.

Before anyone enters the borehole to carry out work at the bottom the sides must be inspected by a competent person from a safety cage fitted with a roof strong enough to withstand falling material or other objects. He must certify it safe to enter (or otherwise). While carrying out this inspection he should remove any loose boulders or stones from the sides. Furthermore, if the borehole is to be unlined except at the top the inspections must be made at frequent intervals while people are working at the bottom.

No descent is to be allowed into an unlined borehole more than 12 hours after the start of excavation or more than 3 hours after completion — these periods may need to be shortened. Also, no person should remain in a borehole for more than 1 hour at a time.

Where strata are unstable, and in every borehole for the first metre, lining tubes must be used. At the completion of excavation and before concreting operations are begun particular care is needed with the extraction of the lining tubes to prevent any overloading of the lifting appliance. Where work is to be carried out within the under-reams the roof of the under-ream must be supported by traditional timbering.

Respiration. At least $1\cdot5\,m^3$/min of respirable air must be provided at the bottom of the borehole for each person employed there. As for shafts, the dangerous gases likely to be encountered are methane, carbon dioxide, possibly carbon monoxide from vehicle exhausts at the surface, nitrous fumes if any blasting has taken place, and possibly ammonia and petrol fumes in contaminated ground from previous occupancy; but perhaps the major danger is from oxygen deficiency. Continuous monitoring for these gases and for oxygen deficiency is most important while anyone is underground and must in the first instance always be carried out by a man wearing appropriate breathing apparatus before anyone is allowed to enter.

Safety equipment and rescue. Those working in shafts and boreholes should wear safety harnesses and safety helmets at all times. Suitable rescue and first aid equipment including a stretcher, breathing apparatus, rescue lines and a winch (or crane) should be readily available for use near the top of each shaft (or borehole). While anyone is employed at the bottom an experienced banksman or rescuer must be on duty at the surface to observe that those at the bottom remain fully conscious and in no difficulty. Where the shaft or borehole is too deep for direct vision some form of communication must be established, such as field telephone, and regular checks made to ensure that those at the bottom are in no difficulty.

In an emergency the person at the surface is responsible for getting the men to the surface as quickly as possible. He should first summon additional help. He should not, nor should anyone else, enter the shaft or borehole without breathing apparatus. Many multiple fatalities have occurred where rescuers have entered confined spaces without breathing apparatus to help colleagues, and have themselves been immediately overcome — especially where the emergency was due to oxygen deficiency.

Scaffolding

In the construction industry perhaps the most common way of providing a place to carry out work at a height is by means of a scaffold. The

main hazards associated with scaffolding are the following
- people falling from the working platforms
- people below the working platform being struck by material falling or being thrown from it
- the scaffold, or part of it, collapsing and throwing people from the working platform; or the collapsed structure crushing people under it or nearby at ground level.
- the collapsed scaffold causing damage to adjacent property or to the structure associated with the scaffold.

All scaffolding should be erected, altered and dismantled by competent and experienced people under competent supervision. A test of competence is that the scaffolder has been trained under the Construction Industry Training Board scheme, and has received the appropriate certificate.

When it has been erected, the scaffold must be inspected at intervals not exceeding seven days by a competent person whose duty it then is to complete the Scaffold Register as required by the Construction (Working Places) Regulations 1966 (*see* p. 14). There is a requirement for the completed scaffold also to be maintained by experienced and competent people.

Tube and fitting scaffolds

Tube and fitting scaffolds are constructed from loose tubes and a variety of fittings; the tubes are clamped together by fittings called couplers. Tubes and fittings are usually made of steel but may be of aluminium alloy. Steel and alloy tubes and fittings should never be used on the same scaffold.

The design and layout of tube and fitting scaffolding are fully described in BS 5973: 1981 *Code of practice for access and working scaffolds and special scaffold structures in steel*.[3] The most common types of tube and fitting scaffold are the putlog or bricklayer's scaffold and the tied independent scaffold.

Putlog scaffolds. The putlog scaffold is mostly used in connection with the construction and maintenance of masonry and brickwork structures and is sometimes referred to as the bricklayer's scaffold. It consists of a single line of standards (uprights) parallel to the face of the building and set about one metre away from it (to accommodate a platform four or five boards wide). Horizontal tubes called ledgers are connected to the standards at vertical intervals of approximately 1·35 m by right-angle (load-bearing) couplers. Putlog tubes span the distance between the ledgers and the building and support the working platform. The flattened blade or spade end of the putlog sits on the brickwork and is built in as the work progresses. The outer end should be fastened to the ledgers by putlog couplers to prevent movement. The decking platform usually consists of scaffold boards 225 mm wide × 38 mm thick and only one working platform may be in use at any one time.

Tied independent scaffolds. An example of a good tied independent scaffold is shown in Fig. 4.5. It consists of a double row of standards each

Fig. 4.5. An example of a good tied independent scaffold (courtesy HSE)

parallel to the face of the associated structure. The inside row should be as close as possible to the structure and the distance between the lines of uprights should accommodate a platform of four or five boards. Sometimes the first row of standards is set away from the building face to allow for one board to be positioned between the building and the line of standards. The lift height (the distance between the ledgers) may vary between 1·8 m and 2·7 m but is generally about 2·0 m. The boards forming the working platforms rest on tubes called transoms which span the distance between the ledgers. These should extend inwards to touch the building face.

A typical independent scaffold with rakers instead of ties at the lower levels is shown in Fig. 4.6.

Falls from scaffolding

Workmen sometimes lose their balance at working platform level because they slip, trip, stumble or on very rare occasions are taken ill. The presence of toe boards and guard rails on a scaffold will generally prevent a fall to a lower level should a workman lose his balance. Toe boards and guard rails should be fitted to the outer side and ends of the working platform and be so secured as to prevent their outward movement — this

Fig. 4.6. An example of a good independent scaffold with rakers instead of ties at the lower levels (courtesy SGB)

is best carried out by positioning them on the inside of the standards. The toe boards (e.g. scaffold boards on edge) should be at least 150 mm high from the decking (i.e. the working platform) and constructed so that no gap exists between the decking and the lower end of the toe board. Guard rails usually consist of steel scaffold tube erected horizontally and fixed to the standards by means of right-angle (load-bearing) steel couplers. They should be located between 910 mm and 1·15 m above the decking and the gap between the top of the toe board and the underside of the guard rail should never exceed 765 mm — the provision of a second guard rail is the usual method of ensuring this. An example of bad scaffolding is shown in Fig. 4.7.

A hazard which might cause a person to slip or trip should of course be identified and eliminated to prevent the possibility of an accident. Areas that are slippery after spillages should be immediately cleaned and sanded as necessary. Chippings (e.g. from bricks) and dust from materials should be swept up and not allowed to accumulate.

Materials deposited on the scaffold platform should be neatly stacked and a clear passageway of at least 430 mm should be maintained between the materials and the edge of the platform. Where it is intended to transport materials along the platform (e.g. in a wheelbarrow) a clear passageway of at least 600 mm is necessary. For the transport of materials along the scaffold platform neither the guard rails nor the toe boards should be removed.

The surface of the platform should be even and without excessive sag caused by the self-weight of the boards, by people moving along it or by the transport of materials (e.g. in barrows). All the boards forming the platform should therefore be of the same thickness and this should be chosen having regard to the intended loading and span. The recommended maximum span for the most commonly used scaffold boards where the platform decking consists of a single layer of boards and the distributed loading is not intended to exceed $3\,kN/m^2$ is 1 m for 32 mm board, 1·5 m for 38 mm board and 2·6 m for 50 mm board. Where a higher loading figure is required the size and spacing of the boards will need to be specially designed.

In addition, each board should be supported by at least three transoms or putlogs. Their ends usually abut and one transom or putlog is required near each end of a board. The transoms or putlogs should be spaced so that the board ends do not overhang by more than four times the thickness of the

Fig. 4.7. An unacceptable scaffold at working platform level with no guard rails or toe boards, a large gap between the boards and inadequate passage width (courtesy HSE)

56   CONSTRUCTION SAFETY HANDBOOK

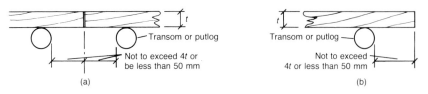

Fig. 4.8. Spacing of transoms or putlogs: (a) at a joint; (b) at the end of a run

board, e.g. 150 mm for a 38 mm board. This will prevent the possibility of the board tipping under load and throwing someone off balance, e.g. a workman walking on to the cantilever or a workman pushing a loaded wheelbarrow with the wheel running on to the area. Also, to ensure that through vibration or other movement a board does not slip off its end support the overhang should be not less than 50 mm (Fig. 4.8).

However, sometimes the boards overlap at the joint and then only one supporting transom or putlog is required. In this case it is necessary to provide bevelled pieces, suitably secured, to overcome the tripping hazard and to facilitate the passage of wheelbarrows (Fig. 4.9).

Workmen may also fall through the scaffold platform if a large enough gap exists or is created. To prevent this the boards should be laid with the minimum of space between adjacent boards. Also, boards should not be removed from working platforms during the life of the scaffold except when they need to be replaced, in which case the replacement should be immediate, or that part of the scaffold that is affected should be fenced off to prevent workmen using it until the replacement has been carried out.

Falling materials

Where there is any possibility of people below being struck by material or tools falling through a gap in the working platform the boards must be laid close boarded, i.e. side by side and end to end without space between the edges of adjacent boards.

As well as preventing people slipping from the outer edges, the toe board has another function, i.e. preventing material and tools falling. Where materials are stored on the scaffold to a height above the 150 mm high toe board either a higher toe board (e.g. two or more above each other) is necessary, or preferably the space between the upper guard rail and the top of the toe board should be filled in with, for example, stout wire mesh frames suspended from the guard rail. They are often referred

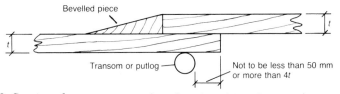

Fig. 4.9. Spacing of a transom or putlog where boards overlap at a joint

to as brick guards and it is of course essential that their outward movement be prevented.

Material, tools or other equipment should never be thrown from the platform but should always be lowered under control in a suitable skip by means of a gin wheel, scaffold jib crane or hoist. This also applies to any component of the scaffold during the dismantling phase of the scaffold structure.

Material or equipment should also be conveyed from a lower level to the working platform by similar means. When lifting or lowering by means of gin wheels or scaffold jib crane it is necessary to ensure that the skip is not accidentally detached from the suspension hook during its travel — only safety hooks should be used. Also, by careful control and visual monitoring of the path of the load throughout its travel, contact with the scaffold and subsequent tipping of the skip should be avoided.

In the construction of scaffolds other methods of preventing people being struck by falling objects include total enclosure — the suspension of sheeting (e.g. canvas sheet) at all the outer edges or the use of protection fans projecting from the scaffold (Fig. 4.10). The decking for fans may be of waterproof sheeting, close-boarded timber or safety netting — where the latter is used it is important for material netting to be laid over the heavier safety netting to prevent small objects falling through the larger mesh.

Scaffold collapse

Partial or total collapse of putlog and tied independent scaffolds may cause major injury or death both to users and to those below the scaffold and considerable damage to adjacent property and to the structure associated with the scaffold.

The scaffolds should be erected in accordance with British Standard BS 5973 and failure to do so in any respect might cause or contribute to collapse. Some examples of poor construction commonly found are

- standards erected out of plumb
- putlog clips used in places where load-bearing couplers should be used
- fittings incorrectly (loosely) tightened
- excessive spacing of standards
- inadequate bracing
- poor foundations, for example standards founded on uncompacted ground, baseplates omitted (only acceptable when standards are pitched on concrete or steel of adequate thickness), lack of suitable sole plates.

However, the main cause of collapse falls into one of two categories: insufficient ties and overloading.

Ties. Ties are required to attach the scaffold structure rigidly to the associated building and they should prevent movement of the framework towards and away from the building. There are two classes of tie, movable and fixed. The use of movable ties is necessary to allow following trades

Fig. 4.10. Use of sheeting and fans to protect those below from falling materials during construction of Natwest Tower, London (courtesy National Westminster Bank and John Mowlem)

Fig. 4.11. A dangerously overloaded scaffold (courtesy HSE)

(plasterers, glaziers, etc.) temporarily to remove a tie (one at a time), in order to carry out their work, but the ties should be replaced as soon as possible. Fixed ties must not be removed. It is essential for the stability of the scaffold that the rules in respect of ties and their spacing as required by BS 5973: 1981[3] are complied with.

Overloading. When a scaffold is in use to facilitate construction, maintenance or demolition of a building or structure, overloading may result from the stacking and storage of material associated with the works on the platform(s)(Fig. 4.11). Generally well-constructed scaffolds will accommodate a degree of overloading but collapse may occur with relatively small overloads where the number and spacing of ties is inadequate and/or one or more of the faults listed on p. 57 are present.

Depositing material on to the platform(s) from a crane or fork-lift truck requires a great deal of care to prevent a heavy landing, and thus a high live load, for which the scaffold is not designed. This is best dealt with by a banksman at landing level directing the crane or fork-lift truck driver regarding the movement of the load. Similarly, a banksman is vitally important for the safe conveyance of loads by crane, and, where the driver does not have a clear view of the whole travel of the load, for preventing collision with the scaffold.

The scaffold platform is sometimes used to stack material from demolition operations and care must be exercised to prevent large pieces of masonry or other material dropping on to the working platform. Also, as the building is progressively reduced in size, it is necessary, especially at the lowest levels, to ensure that the required numbers of ties are maintained or that the scaffold is stabilised by means of rakers.

It is also fairly common for collapse or partial collapse to occur during dismantling. This usually happens when material (tubes, boards and couplers) from the higher areas of the scaffold is stacked at the first or second lift level to await a lorry to take it away. The material is stacked over a relatively small area of the scaffold, resulting in a massive overload which the remaining structure is simply unable to carry. Preferably the material should be lowered to the ground and stacked to await removal, or if this is not possible and it has to be stored at the lower levels, the scaffold should first be adequately strengthened. Also, dismantling should be evenly carried out in horizontal layers to the full width of the scaffold and not in vertical sections.

Other causes. Other causes of partial collapse are

- in putlog scaffolds, the dislodgement of the spade end of a putlog causing the platform to fall
- the failure of a scaffold board(s)
- collision of motor vehicles with the structure.

In the case of putlog scaffolds the spade ends of the putlogs must rest on or in the brickwork to a depth of at least 75 mm. This will prevent accidental displacement caused by vibration and other scaffold movement.

Scaffold boards should be carefully examined immediately before their use in a working platform. They should be free from unacceptable faults such as large knots, knot clusters or large splits. However, perhaps the most common reason for the failure of scaffold boards is reduced strength owing to a compression fracture (of the wood grain) from previous careless use. The fractures are often caused when boards are dropped from a height, especially if they land on a sharp edge such as the edge of a concrete raft or on the edges of broken bricks on the site, and when boards are left lying about the site and vehicles run over them. Compression fractures greatly reduce the strength of a board and are very difficult to identify even by experienced examiners. Also, boards should never exceed the spans recommended on page 55.

Unless the scaffold is constructed in an area where little or no traffic is anticipated the standards should be protected at their base by fenders (e.g. large timber baulks laid around the outer edges of the structure) to prevent vehicles colliding with the structure.

Proprietary scaffolds

Independent tied scaffolds may also be constructed from proprietary prefabricated units of which several systems are available. The safety precautions for tube and fitting scaffolds apply also to proprietary systems; they should be plumb and on firm bases, with all joints secure, and they should not be overloaded. Although proprietary scaffolds may not require the same standard of bracing as loose tube and fitting scaffolds they all require the same standard of tying. When cranes are working in the vicinity, care should be taken to prevent snagging the scaffold because, in some types of proprietary system, the joints have no resistance to an upward thrust.

Tower scaffolds

Tower scaffolds may be of the stationary or mobile type. They are usually rectangular or square towers constructed from loose steel tubes and fittings or proprietary types comprising prefabricated frame sections with special joint arrangements making assembly simple and speedy (Fig. 4.12). They are also often made from aluminium alloy as well as steel. In all cases they provide one working platform. To prevent people falling from the working platform, guard rails and toe boards are required as for the common putlog or independent tied scaffolds. Also, similar transom spacing, board spans and board qualities are required.

With mobile towers a common accident is for the scaffold to move inadvertently while someone is on the platform, and a wheel(s) hits an obstruction with the result that either the scaffold overturns or tips sufficiently violently to throw the person(s) from the platform. To prevent this it is essential that all the wheels (castors) are provided with efficient locking devices and that these are engaged while people are using the platform. Also, the wheels must be positively secured to the uprights to prevent them falling off. A similar type of accident results from the tower being deliberately moved at the base (to a new working area) while the platform remains occupied. This action should never be carried out — no one should occupy the platform while the tower is being moved. Also, the towers should only be used on firm and reasonably level ground.

Towers may also overturn where an excessive horizontal force is applied — usually at the working platform by the user, but the force may be due to wind load in a location outside a building. A general rule for ensuring stability for most operations is that the height to smallest base dimension ratio must not exceed the following in respect of steel towers

- 4:1 for stationary towers used inside a building
- 3·5:1 for stationary towers used outside
- 3·5:1 for mobile towers used inside a building
- 3:1 for mobile towers used outside.

The height is measured from the floor to the working platform level and the least base dimension is the width from centre to centre, of the shortest side. Where outriggers are used the width over the shortest side of the outrigger base is used. Where aluminium towers are used the above ratios would be reduced.

A particularly common accident with tower scaffolds occurs where the operative(s) wishes to reach a work area substantially higher than the

Fig. 4.12. A good example of a mobile work platform (courtesy HSE)

base-to-height ratio for the platform level will allow. In order to overcome this difficulty, frequently operatives erect a ladder on the platform, usually with the foot of the stiles hard against the toe board. Sometimes one person will 'foot' the ladder while the other starts to climb to the working area. Inevitably the horizontal reactive force at the bottom of the ladder due to the weight of the ladder and the person (plus tools or material) climbing produces a much greater overturning moment than the scaffold's natural (self-weight) righting moment and the scaffold overturns, usually with disastrous results. This method of work — placing a ladder on the working platform of a tower scaffold — should never be used.

Birdcage scaffolds

Birdcage scaffolds are intended for light work such as decorating, plastering or fixing lighting, or for light maintenance work within buildings with large floor areas such as factories, cinemas and churches. They consist of parallel rows of standards at regular intervals evenly spaced in both directions — not exceeding 2·5 m centres. They are connected using right-angle couplers to ledgers and transoms at each lift height, which is generally about 2 m. The top lift only is boarded out to form the working platform, and here the transom spacing must be such that the span of the boards is appropriate to their thickness (*see* p. 55). Very often birdcage scaffolds are erected to cover the whole of the floor area of the part of the building where they are used. If the platform is then constructed so that no, or only a very small, gap exists at the edges, toe boards and guard rails are unnecessary but if the gap is over 25 mm and people below are likely to be struck by falling material then toe boards should be used. If the gap is such that someone might fall through the opening, guard rails are also required.

The most common accident with this type of scaffold is either complete or partial collapse (Fig. 4.13). It is usually caused by very excessive overloading coupled with various elements of poor construction (e.g. standards not plumb, use of incorrect, non-load-bearing couplers). However, the most common fault is the lack of adequate bracing in two directions to prevent excessive sway and ultimate collapse. Additional measures to prevent collapse are adequate tying and the extension of the transoms and ledgers to butt against the sides of the building (around all the sides of the scaffold). Ties may be of the box type around suitable columns within the building or reveal at window openings. One such restraint point (either a tie or a butting tube) should be provided for each 40 m^2 of vertical face and no standard should be more than six tubes away from a restraint point.

Fig. 4.13. Collapse of birdcage scaffold due to inadequate bracing (courtesy HSE)

Slung scaffolds

Slung scaffolds are working platforms suspended by means of lifting gear, wire ropes, chains or rigid members (usually scaffold tubes) at a fixed distance below structural elements of a building (e.g. roof trusses or other suitable overhead load-bearing members). The platform is usually supported on scaffold tube ledgers and transoms which should be connected by right-angle couplers. The transoms should be spaced so that the span of the boards does not exceed that appropriate for their thickness. As for all scaffolds the platform should be close boarded and provided with suitable guard rails and toe boards at the outer edges. Where two transoms are used to support the joint of two boards their distance apart should allow for an overhang of at least 50 mm but not greater than four times the thickness of each board. This type of scaffold should always be specially designed and the suspension points for the platform supports should not exceed 2·5 m centres.

A major hazard particular to this type of scaffold is the failure of the suspension arrangement, either the suspension gear or the anchorage. In order to avoid or at least substantially to reduce the likelihood of an accident it is necessary to adhere to the following points.

- Where wire ropes or chains are used they should be adequate for the purpose and should have been tested. Scaffold lashings should never be used.
- A sufficient number of suspension points should be provided to prevent the platform tipping if one of the suspension ropes etc. fails (e.g. at least five points are needed for a circular platform and six for a rectangular platform).
- The wire rope etc. suspenders should be kept vertical as far as possible — if they are not this must be taken into account in the design.
- The wire ropes etc. should be secured to the platform ledgers as close as possible to their junctions with the transoms.
- At the platform end, wire rope suspenders should be secured to the ledgers by two full turns around the ledger and the end held back to the suspension wire by three bulldog grips, or by means of a suitable shackle when the end of the rope terminates in a tested eye.
- At its upper end the rope etc. should similarly take two round turns of the suspension member and the tail should be secured to the running rope by three bulldog grips.
- Where the wire rope at the upper suspension point is taken around sharp edges, the rope should be protected at all such high stress raising points by means of appropriate packing.
- The suspension ropes should have a factor of safety against failure of at least 6.
- The point of suspension and member from which the scaffold is suspended (e.g. members of a roof truss) should have a factor of safety against failure under total maximum load conditions of at least 2.

- Where suspension chains are used both ends should be secured by shackles.
- Where scaffold tubes are used as suspenders only right-angle couplers should be used and check couplers should be incorporated to protect against possible slippage.
- Vertical axial joints in tubes should not rely solely on sleeve couplers or joint pins. The joint should be made with a lapping tube and a suitable number of right-angle couplers.
- Material and equipment should be spread evenly over the platform and heavy point loads should be avoided. Workmen using the platforms should not congregate in any one bay.
- Guy ropes should be used to minimise platform sway.

Suspended scaffolds

Suspended scaffolds are sometimes referred to as cradles and are used for light work such as painting and repair work on the outside of buildings. The permanent type of power-operated suspended scaffold provided on some buildings is not dealt with here.

The cradles are suspended by fibre or wire ropes and are capable of being raised or lowered by manual or power operation. They may be of fixed type or be capable of travelling horizontally on a runway as well as being raised and lowered. The most common accident occurrence is the scaffold falling because of failure of the suspension ropes or the outriggers or, with the travelling type, running off the end of the track. Suspension ropes should not be used if they have suffered undue wear or damage that affects their strength. They should be properly reeved on to winding drums or winches (where used) over all pulleys and be correctly secured at anchorage points. Where the winches are of the drum type the length of the suspension rope should be such that at the cradle's lowest point there should be at least two dead turns of rope on the drum — to alert the operatives the last few metres of rope should be distinctly coloured. The rope anchorage on the drum should also be capable of sustaining the rated load of the winch. Also, winches used in connection with the scaffolds should be provided with a brake which comes into operation when the operating handle is released.

Outriggers may be of steel joists, timber poles or sometimes suitably braced steel scaffold tubes. They should be

- of sufficient strength for the maximum load to which they will be subjected
- project beyond the face of the building just far enough to ensure that the suspended cradle will not foul on any parts of the building during its upward or downward travel
- either bolted down at their inner (tail) end or securely counterweighted to prevent overturning under maximum load conditions; the total load includes dead load, live load and an impact factor (25% for power-operated and 10% for manually operated scaffolds); the counterweights where used must be properly attached to

the tail end of the outriggers — loose bags of sand or rolls of roofing felt are not suitable methods of counterweighting; the counterweighting should provide a resisting moment of at least three times the maximum overturning moment
- kept as nearly horizontal as possible but never with the outer end lower than the tail end
- spaced reasonably close together but suitable for the maximum loads to be carried.

Where a runway is incorporated to facilitate sideways (horizontal) movement of the cradle it must be securely lashed or shackled to the outriggers. The jockey wheels are prevented from running off the end of the runway beam by securely fixed end stops at each end of the beam.

Ladders

Access to many scaffolds is provided by ladders — usually timber ladders. Also, the ladders themselves are very often used as working platforms for the performance of light work, e.g. painting or inspection. Regrettably each year many serious accidents result from the misuse of ladders. Accidents occur because

- ladders slip when users are climbing or working from them; over 50% of all ladder accidents may be attributed to this cause
- users slip or miss their footing while climbing
- users overbalance when carrying materials or tools
- when defective ladders are used they fracture under the weight of the user.

When used for access, ladders should be securely tied near their upper end — preferably by each stile to the ledger at platform level. They should also stand evenly on their stiles, on a level and firm surface — never on loose bricks and never carried by a rung resting on a plank across a depression (e.g. a trench) in the ground. The rungs and users' footwear should be kept clean and free from slippery mud, for example. A particularly long ladder should be secured to prevent undue sway. At scaffold platform level the ladder must extend at least 1·05 m above the landing place (platform level) to provide a handhold while the user steps on or off. If this is not possible a suitable alternative handhold must be provided. There should also be sufficient space at each rung to provide a good foothold.

When carrying materials and tools invariably the user is climbing with only one hand on the stile. Alternative methods of lifting the material and tools should always be considered.

Where work is to be performed from the ladder, where the ladder cannot be tied at the top or adequately secured at the bottom to prevent movement, the ladder should be footed. The ladder should extend beyond the working place by at least 1·05 m unless other adequate handhold is available. Many accidents are caused by over-reaching, causing the ladder to slip sideways — if the ladder is not secured at the top the user must exercise great caution when reaching sideways. Generally, ladders should be placed at an angle of 75° to the horizontal, i.e. 1 m out for every 4 m of vertical height when used for both access and working purposes.

Timber ladders should be carefully handled — as with scaffold boards, they should not be dropped or thrown from heights and they should not be left on the ground where vehicles may drive over them. They should be stored in ladder racks and not hung by their stiles. Makeshift ladders should not be used, particularly those where the rungs depend solely for support on nails, spikes or similar fixings.

Ladders with missing or defective rungs should never be used and where wire tie rods are used the ladder should be erected so that the rods are below the rungs. Metal ladders and timber ladders with metal reinforcements should never be used in the vicinity of electrical apparatus or equipment.

Ladders should be regularly inspected to identify defects and timber ladders should not be painted since this may hide defects — transparent varnish or linseed oil may be used as a preservative.

Falsework

Falsework is defined as a temporary structure used to support a permanent structure during its construction and until it becomes self-supporting. Falsework may be required to support steel and timber frameworks and masonry arches as well as in situ and precast concrete construction. The design and construction of falsework is covered by BS 5975.[4]

Falsework is often constructed from conventional loose steel scaffold tubes and fittings together with adjustable telescopic props and forkheads (sometimes flat heads) which carry the main timber bearers on which the formwork or deck is built. Secondary beams may be interposed between the main bearers and the formwork. Prefabricated steel frames, military trestles, towers, steel girders and standard steel sections are employed for more specialised work and generally where the load to be supported is particularly heavy.

The total or even partial collapse of falsework may lead to serious accidents for those on the structure who are thrown from or fall from their place of work (Fig. 4.14).

More often than not collapse occurs when the structure is being loaded, e.g. during the placing of concrete when the number of workers involved in the operations is high. Where other work is being undertaken below the loading level the people involved are also at considerable risk from falling materials, i.e. both from the falsework and from the supported permanent works.

Accidents also occur without the falsework collapsing. These are the result of people slipping or falling or both from the structure. In addition to their major supporting roles the falsework and the formwork provide a place of work, and normal edge protection requirements are necessary at all open sides and holes in the decking, as well as properly constructed access to all places where work is to be carried out (Fig. 4.15). The collapse of falsework can be attributed to inadequate design, poor construction or a combination of the two.

Fig. 4.14. Collapse of falsework for Loddon Bridge, Berkshire (courtesy HSE)

Design

All except the very simplest falsework should be designed in accordance with recognised engineering principles and clear drawings, sketches and specifications should be produced for use by the site personnel so that the structure can be built in accordance with the designer's intentions.

For fairly simple repetitive applications standard solutions may be used. Some examples are given in BS 5975 *Code of practice for falsework*.[4] However, when the standard solutions are used they must be developed by means of proper engineering design methods and the users on site must always be provided with drawing(s) or sketch(es) and if necessary a specification to ensure that construction is carried out in accordance with the standard solution.

Deficiencies in falsework design may arise from

- failure correctly to estimate the type and extent of the loading
- inadequate foundations
- incorrect choice or use of materials
- lack of provision for lateral stability.

Failures may also occur because of design errors which, given the temporary nature of falsework, are prone to occur if checking is neither thorough nor systematic.

Loading. Falsework collapse has occurred because the designer has

HAZARDS OF CONSTRUCTION AND THEIR PREVENTION

failed to take into account all the loads applied to the structure. In addition to the self-weight of the falsework and formwork, the ancillary temporary works connected to the falsework such as hoist towers, loading platforms and access ramps, the following are common examples of the imposed loads that must be taken into consideration when they apply

- the self-weight of the permanent works to be supported
- dead loads from static plant such as welding sets and compressors
- dead load due to stored materials, i.e. reinforcement or timber
- impact in placing the permanent works (e.g. units lowered by crane on to the supporting structure); where wet concrete is being placed, the impact when tipped from a barrow, a dumper, a skip or a concrete pump pipeline
- where wet concrete is to be pumped through a pipeline to the placement position both the dead load of the pipeline and the horizontal load due to the impulse effect
- the surge effect of wet concrete due to the use of poker vibrators
- those due to the hydrostatic-type pressure of fresh wet concrete on the faces of the formwork; the effect is more complex if the soffit is sloping
- those due to the expansion of concrete during hydration
- rolling loads from moving plant such as dumpers and barrows, including the effect of braking with mechanically propelled vehicles
- moving load due to operatives and an allowance for small plant and tools that they use in the course of their work
- those due to post-tensioning of the permanent works

Fig. 4.15. Safe working platforms on falsework for bridge piers for Dartford River Crossing (courtesy Cementation Cleveland Dartford Crossing)

- wind load on the falsework, the formwork and the completed part of the permanent works carried by the falsework; also, as necessary a snow load
- where the falsework is founded in a water course the dynamic action of the flowing water on the supports; consideration should also be given to the impact of floating debris on the supports and to the load imposed by the flowing water impinging on debris trapped by the supports; sometimes the falsework is erected in a dry watercourse but flash floods should be anticipated and the above load considerations applied.

Foundations. Many falsework structures are founded on railway sleepers or other types of sole plate and the designer should specify the need to ensure that the underlying ground is well consolidated and that any voids are filled with concrete.

The stability of the foundations may be seriously affected by scour from storm-water. If this is likely there may be a need to provide cut-off drains and to place the sole plates on surface blinding. Non-cohesive soils are particularly vulnerable to the effect of water and a high water table may considerably reduce the ground bearing capacity. Where uncompacted they are also susceptible to consolidation under the effect of vibration, e.g. of the wet concrete — which may cause differential settlement. When falsework is erected in dry watercourses a more substantial concrete foundation may be necessary to prevent scour caused by flash floods.

In the case of timber sole plates, the designer should anticipate and allow for an eccentricity of 25 mm for the uprights on the centre line of the sole plates. Where the sole plates are founded on sloping ground the falsework uprights and base plates should be built on wedge-shaped timber packing with cleats immediately below the wedges, wedges and cleats properly secured to the sole plates. Thrust blocks at the base of the slope are required to prevent the movement of the sole plates down the slope.

At the design stage it is necessary to take into account the intended loading pattern to avoid differential settlement of the foundations. Such settlement would cause a redistribution of the loads on the falsework structure and might lead to an overload.

Where the falsework is supported on the permanent works such as in multi-storey building construction the falsework foundation design should be based on the strength of that part of the permanent structure that is intended to carry the falsework. The strength will be that at the time of loading and not necessarily the ultimate strength.

Materials. For falsework applications the materials used are often not new, but may have been used many times, especially for repetitive work of a similar nature. Materials may have been used for other types of work and may have been subject to abuse by overloading, rough handling or poor maintenance. The designer must be aware of the limitations of the material to be used for falsework under consideration and should adjust his permissible stresses accordingly. A particularly common reason for the need to reduce allowable stresses is the reduction in cross-sectional area of steel components due to the effect of corrosion. Where second-hand material is to be used it should first be inspected. Where cross-sections of components are severely reduced by corrosion or by previously drilled holes or where they are excessively damaged, bent or twisted, they should be rejected.

Stability. One of the main causes of collapse is the lack of sufficient lateral stability. It is important to prevent initial movement of a falsework system towards a collapse condition. The forces required to prevent initial sway are relatively small but when sway is induced, the inertia of the supported load and increasing eccentricities often make collapse inevitable.

The various horizontal forces due to some of the loadings identified under loading must be resisted by an effective bracing or restraint system. In addition to any bracing system, the designer should where possible require the falsework to be tied back to completed parts of the permanent structure or be firmly anchored to a suitable foundation.

Where an accurate assessment of the horizontal loads cannot be made horizontal forces equivalent to $2\frac{1}{2}\%$ of the applied vertical loads, considered as acting at the points of contact between the vertical loads and the supporting falsework, must be catered for as advised by BS 5975 *Code of Practice for Falsework*.[4]

Forces on mobile falsework. Some falsework or parts of falsework are moved on skids or wheels to another position for reuse. The designer must take into account the additional forces that may be imposed on the structure because of this. The need to overcome uneven track or ground conditions may subject the structure to excessive twisting or bending and this may also occur when manoeuvring it into its next position.

Fig. 4.16. A good example of temporary beam grillage showing correctly placed web stiffeners (courtesy HSE)

Fig. 4.17. Timber baulks provided as fendering to protect falsework from vehicle impact (courtesy HSE)

Buckling. It is important to detail all points of load transfer (e.g. from timber bearer to forkhead) in the knowledge that slight misalignments in erection can cause buckling and progressive failure.

When steel beams form part of the structure it is important that at all load transfer points the webs are checked for their ability to resist buckling (Fig. 4.16). Where calculations are not produced which satisfy this requirement web-stiffeners must be provided at all such points. Failure to do so will almost certainly lead to collapse.

Traffic access. It is quite often necessary to allow vehicular traffic to pass through the falsework structure, e.g. where it straddles a roadway. To prevent vehicles colliding with the framework, a suitable fendering system should be specified (Fig. 4.17). This should include a lead-in phase and not just comprise the immediate protection of the falsework.

Design brief. To produce a suitable solution for the falsework requirement the designer must be provided with a brief. This will include such information as

- the self-weight of the permanent structure to be supported
- the falsework components to be used
- the access requirements to and through the structure
- data from the soils investigation

- information in respect of watercourses and sources
- the method of construction and the loading sequence
- the plant to be used and the method of placement
- the storage requirements e.g. at deck level
- in multi-storey construction the load-bearing capacity of the floors.

Construction

It is of paramount importance that the construction of the falsework follow exactly that shown on the drawings and in the specification produced by the designer. No alteration whatsoever should be made without consultation with and approval of the person responsible for the design. It is also important that activities which have not been identified in the design brief but which may affect the stability of the falsework should not be carried out without clearance from the designer, e.g. excavations in the vicinity of the structure.

Failure to observe basic details of good construction will often be the cause of a partial or total collapse. Some examples are now given.

Materials. All materials should be inspected before use and any not of the required quality should be removed. Where scaffolding components are to be incorporated only load-bearing couplers should be used, and putlog clips should be taken away from the storage area.

Fig. 4.18. Poor foundation for falsework: vertical members should be located centrally on their sole plates (courtesy HSE)

Foundations. The area beneath the sole plates should be well consolidated and where required the blinding should be laid or the concrete foundation constructed to the specified depth and area. The vertical members of the structure should be located centrally (laterally) on the sole plates of base grillage and none should be within 300 mm of the ends (not as in Fig. 4.18).

Adjustable steel props and forkheads. Only props of the correct size should be used. Load is transferred from one part of the prop to the other through a high-tensile pin and this should never be replaced by a makeshift pin, e.g. a reinforcement bar.

Props should not be bent and should be erected as plumb as possible but never more than $1\frac{1}{2}°$ out of the vertical, i.e. not more than 25 mm in 1 m. They should be positioned centrally underneath the member to be supported — any eccentricity should never be greater than 25 mm. At the top of the prop the supported timber beam should be square to the plate and secured in position by nails.

With adjustable forkheads the supported bearers should be positioned to avoid any eccentricity greater than 25 mm on the forkhead. They should be rotated to centralise the bearers over the tubular stem and matched wedges should be used to maintain the bearers in place. In addition the wedges should be secured by nails through the holes in the sides of the forkheads. The bearers should preferably be butt jointed and the joint should be within 15 mm of the centre. Where a beam terminates in a forkhead it should extend at least 50 mm beyond the centre point.

Where lapped joints are used for the bearers they should not span more than three supporters. Over-extension of the screw jack supporting the forkhead must be avoided. This must not be more than 300 mm without lacing (horizontal bracing) at right angles to the bearers, as near to the top as practicable, and with diagonal bracing in both directions at every sixth jack. Where adjustable base plates are incorporated similar bracing arrangements are required where the jack extension exceeds 300 mm.

Scaffold support components. The uprights should be centralised on the sole plates as for props and they should be erected plumb within a tolerance of 15 mm over a height of 2 m, subject to a maximum of 25 mm over the full height of the structure. Only load-bearing couplers should be used and these must be properly tightened.

For fabricated steel floor centres the correct spacing must be ensured and full bearing achieved at each end of individual centres on the end bearers. They have little or no resistance to horizontal forces and an adequate bracing system is essential before concreting is carried out.

Loading. Where precast concrete or prefabricated units are lowered on to the falsework a banksman should control the lowering operation and ensure that shock loading is prevented as far as possible. Also, the dragging of units into final positions should be avoided.

On the deck, materials should only be stored in the areas designated by the designer and should not exceed the permitted load. Only plant specified in the design brief should be allowed on the structure unless clearance is received from the designer.

The placing of in situ concrete should be carried out in the sequence agreed by the designer. It may be taken to the working position by crane and skip, by barrow or dumper, or be piped or taken by mixer truck. The free fall should be limited to a maximum of 1 m for specially designed falsework and to 500 mm where a standard solution has been adopted. Also, heaping of the concrete within a small area should be avoided, e.g. in an area of 1 m^2 the height of the heap above the formwork surface should not be more than three times the depth of the slab unless special allowance has been made in the design.

Checking. Immediately before loading the falsework should be examined and checked by an experienced and responsible person to ensure that it has been constructed properly and in accordance with the drawing(s) and specification (where appropriate). Only when he is satisfied with the whole of the falsework should he give permission to load.

Also, throughout the loading period a careful watch should be kept to identify any structural distress and the necessary action should be taken at the first sign of failure.

Temporary works co-ordinator. British Standard 5975: 1982 recommends the appointment and use of a temporary works co-ordinator. Briefly it is his responsibility to co-ordinate all the falsework activities such as the provision of a design brief, the production of a satisfactory design and the required drawing(s) specification and any other information, the receipt on site by the appropriate authority of the design, ensuring that checks are made throughout the construction phase and that any changes in materials or operations are trasmitted back to the designer for consideration before they are used or implemented, supervising the final check and issuing formal permission to load when the checking has proved satisfactory. He would also be responsible for implementing the inspection during the loading stage and for advising on the action should the structure exhibit any sign of distress at this stage, and finally he would issue formal permission to strike the falsework when the permanent works had attained sufficient strength to be self-supporting.

Erection of structural framework

The most serious accidents that occur during the erection of structural framework are due to

- erectors falling from heights when at their places of work or going to and returning from them
- people at lower levels being struck by tools or materials falling or being thrown down
- the uncovenanted collapse of the whole or part of the framework causing men to fall or striking those at lower levels.

The prevention of accidents depends to a great extent on

- the designer having a good knowledge of safe erection procedures and taking these into account in the detailed design — in particular the need for temporary bracing or other measures to ensure the

stability of parts (or the whole) of the framework should be clearly shown or stated on the drawings and/or in the specification; the specification may also be used to advise on any special sequence of erection necessary to ensure stability of the structure
- good communication between the designer and the erection contractor, for example to ensure that any measures to be used by the contractor that may put extra loading on the uncompleted structure (e.g. by the storage of material or the provision of temporary working platforms) or any proposed design modifications are discussed and approved by the designer
- the production of and adherence to a method statement, which must be kept on site, by the erection contractor which should include detailed requirements for
  - the sequence of erection including aligning and levelling procedures
  - ground level assembly work prior to lifting
  - slinging and lifting
  - temporary bracing and support
  - access
  - safe working places and personal protection including safety nets, belts, harnesses and lanyards
  - the plant to be used including inspection and maintenance programmes
  - stacking and storing arrangements
- good management organisation and control and good site discipline to ensure adherence to the erection plan and the maintenance of safety features (e.g. plant, working platforms and ladders); the person in direct charge of operations on the site should be provided with a copy of the method statement and he must have a thorough understanding of its contents
- the installation of permanent stair access and the structural floor system as early as possible.

Preventing workmen from falling

Workmen can be prevented from falling from the structure by the provision of safe access to and egress from all places of work together with safe working places for positioning structural members, bolting up and other operations. However, accidents during framework erection may be minimised by reducing the need to work at heights by pre-assembling sections on the ground.

Safe access. For *vertical access* men should not be permitted to climb columns unaided or to slide down them. Vertical access and egress may be by means of ladders which may be inclined when serving single-storey heights. Inclined ladders should have wire scaffold lashings or rope lashings attached to the stiles at the upper ends and these should be fixed to a convenient beam to prevent movement. Where the lashings are taken around sharp beam corners, packing pieces should be inserted to prevent high stress points. Before columns are erected vertical ladders may be securely fixed to them; with steel I-section columns, flats (similar to rectangular section ladder rungs) may be pre-welded between flanges near the top end of a column or column length. The flats serve as the last few ladder rungs and may be used as lanyard anchor points for safety belts or harnesses when the top of the ladder serves as a short-term working place, e.g. for the erector to receive and position an incoming beam.

All ladders should be positioned to allow sufficient space at the back of the rungs for a firm footing on the ladder and for the toes of footwear to protrude. In the case of square or rectangular concrete columns timber distance pieces will be necessary between column and ladder stiles.

For *horizontal access* steel erectors should wherever possible use properly fixed lightweight staging or scaffold boards. The staging must have adequate bearing, be properly and evenly supported and secured, and the span must not be excessive for the load to be carried. The walkways should have a minimum width of 430 mm and be provided with suitable guard rails and toe boards along each side where a fall of more than 2 m is possible. The fixing and moving of horizontal access ways should be carried out from a safe position or by erectors wearing safety harnesses with lanyards securely anchored.

A commonly used method of moving across framed structures is along the beams themselves. *Beam walking* along the top surface of concrete beams or flanges of steel beams is unacceptable and should not be allowed — it is very dangerous and has led to many serious accidents.

One method of crossing is by straddling the beam. It should only be used for I-section beams. The erector sits astride the top flange with his feet (heel and toe) resting on either bottom flange. The erector then grips either side of the top flange with his hands and moves along, maintaining this sitting position. For this method the depth of beam must be suitable to allow the sitting position described and there must be no obstruction such as cross beams which necessitate a change in position. The method should only be used for access of short duration.

Where straddling is not convenient and access is of short duration walking along the bottom flange on one side of the beam may be used. The top flange should not be higher than the erector's waist when he stands on the bottom flange, and the top flange width must be such that he can comfortably reach across and obtain a secure handhold. He is then able to move sideways along the bottom flange.

Means of access that also provide safe working places are

- power-operated mobile work platforms (Fig. 4.19)
- man-riding skips
- tower scaffolds
- common scaffolds.

For power-operated mobile work platforms a firm base is required; the platform stands at ground level, normally on the already installed concrete sub-floor. Man-riding skips are suspended from a crane or winch, and controlled by a deadman's handle, i.e. the brake is applied when the

Fig. 4.19. Telescopic hydraulic work platform in use for steel erection (courtesy HSE)

control level or switch is not being held in the operating position. Power lowering should also be provided. The skip should be not less than 910 mm deep and be provided with an anti-spin device. Consideration should be given to securing the men in the skip by means of safety belts or harnesses attached by lanyards to a secure place on the skip. For tower scaffolds a firm level base is necessary — again these are generally used on the already installed concrete sub-floor or in a multi-storey building on the floor beams when these have been positioned.

Safe working places. Where a good deal of bolting or welding at height is necessary a properly constructed working platform with guard rails and toe boards may be provided, preferably fixed at the required position on the structural member (e.g. a column) before its erection. Although the Construction (Working Places) Regulation 1966 do not require guard rails and toe boards to be fitted to such platforms which are at least 800 mm wide, if their use (at any one position) is for such a short period of time to make their provision unreasonable, it is safer not to take advantage of this relaxation.

Some types of lightweight prefabricated platform with guard rails and toe boards are constructed to be lifted up and hooked over structural sections. Other methods of providing safe working places at height include

- ladders for work of short duration: where these are attached to columns a safety belt or harness should be worn and the lanyard should be attached to a secure place (*see* p. 79)

- tower scaffolds, useful for single-storey work or for work from completed floors of multi-storey frames (Fig. 4.20)
- power-operated mobile work platforms which could stand on completed concrete slabs or floors, provided that the floor is of adequate strength
- man-riding skips, suspended from a crane or winch, which are useful (similarly to mobile platforms) for bolting-up work, purlin erection, truss positioning and sag bar erection (*see also* Fig. 4.34); this type of plant makes it unnecessary for erectors to walk up rafter backs to position purlins — a very dangerous operation
- sheeting rail cradles, consisting of two small working platforms with guard rails and toe boards separated by two supporting beams, a bay-width apart; lifting points with shackles provide four-point suspension by a chain sling for lifting and lowering by crane; a rail carrier is incorporated in the cradle, and with one man on each platform, sheeting rails can be erected for the full height of the structure bay by bay.

Safety belts and harnesses (*see also* p. 216). Where working platforms cannot reasonably be provided, personnel should wear safety belts or

Fig. 4.20. Tower scaffold access for frame erection (courtesy HSE)

Fig. 4.21. Sala Block fall arrest device (courtesy Barrow Hepburn Ltd)

harnesses, attached by a lanyard to a secure place. This may be to pre-welded flats, or to the eye bolt of a girder grip device, e.g. the Manlock.

Choke-hitching of lanyards around structural sections is not recommended because of the very high stresses, sufficient to cause failure, that occur in the lanyard in the event of a fall. An alternative is a tested wire rope loop of minimum diameter 8 mm with thimble eyes at each end. The loop is passed loosely around the structural section and the two thimble eyes are threaded into a carabiner shackle attached to the lanyard. The lanyard anchorages should always be above the user to limit any fall distance. In any case, the fall height should not exceed 0·6 m for belts and 2·0 m for harnesses.

For the erection of masts and towers which involves a good deal of movement about the structural frame and where other methods of providing safe working platforms are impractical, inertia reels (Fig. 4.21) or taut steel wire running lines with a safety harness may be used (Fig. 4.22). The structural framework and the anchorages must be sufficiently strong, however, to resist the forces generated by an erector falling.

Safety nets. In some fairly large structures consideration should be given to the use of safety nets as an added precaution to prevent or minimise

HAZARDS OF CONSTRUCTION AND THEIR PREVENTION 81

Fig. 4.22. Ladderlock fall arrest device (courtesy Barrow Hepburn Ltd)

injury in the event of someone falling, but they must never be used as an alternative to the proper working platforms and places described above. A decision should be made during the design stage and suitable fixing brackets should be fitted to the beams or columns before erection. Nets should be provided and used in accordance with the latest version of British Standard *Code of Practice* 93.[5] They must be slung as close as possible to the working areas but never lower than 6 m below them. Good access should also be provided to the fixing points and all measures should be taken to reduce the dangers involved in fitting the nets.

Use of cranes for erection purposes. In addition to losing their balance and falling from a height, erectors can also be knocked down from their working place. This risk can be minimized by the careful control and handling of structural elements when they are being lifted into position. The care and expertise of the crane driver is an essential element: if he cannot see the load for the full extent of its travel he must be guided by a banksman, and the communication between the banksman and driver must be of a good standard. The crane driver must carefully observe not only the travel of the load but also the travel of the jib to avoid colliding with the structure — which might throw an erector off balance. Tail ropes should be fitted to components and handled to eliminate unnecessary swing and the load should be moved only slowly. Components to be lifted must be properly slung to prevent sections falling and should have packing pieces at the corners; moreover the crane support should not be released before the component is properly secured. The erectors themselves must be occupying a safe place either on a properly constructed platform or, where straddling a beam, standing on a beam or standing on the rungs of a ladder, they should be wearing a safety belt or harness with the lanyard effectively anchored to the structure.

Care must be exercised in slinging and transporting materials and components by crane or winch to prevent their becoming loose and falling. Particular care is needed in releasing slings and the use of remote-release shackles should be considered. Where a pole (stick) is used for lifting purposes it should be erected on an adequate foundation to avoid settlement, and should be properly guyed, with the ropes not fouling the framework.

The practice of using more than one crane to lift one component or a single assembly should be avoided wherever possible. However, sometimes *tandem or multiple lifting* has to be resorted to. In this case

- only one person must control the lifting operation and he must be in direct communication with the crane drivers; he must be competent and experienced in tandem lifting
- all involved in the lift must be experienced and fully briefed about their role and the method of carrying out the lift
- where for the lifting operation the cranes are required to move while carrying the load the paths they are to follow must be clearly marked; only the person who is directing the operation may give instructions to the drivers and hence control the movement of the cranes
- the safe working load of each crane for the required jib length and operating radius should be at least 25% in excess of its calculated share of the load to compensate for load shedding from one to the other; this may be caused by uneven ground, load swinging, variation in the relative movement between each crane, or differential bounce effect as the load is moved along uneven ground.

Unless great care is taken in the planning and execution of this dangerous manoeuvre, there is a good possibility that the load will be dropped due to the overturning or overloading of one of the cranes involved (Fig. 4.23).

Preventing tools and materials falling

Wherever possible work should be programmed to avoid the need for workers to be underneath areas of higher level working. Barriers and warning notices prohibiting access should be erected at the approaches to all danger areas. Where it is not possible to exclude work activity below the erectors the people below must be provided with adequate overhead protection in the form of material nets, sheets, fans or some other form of structure to give protection from falling objects.

HAZARDS OF CONSTRUCTION AND THEIR PREVENTION 83

Fig. 4.23. Badly executed tandem lift (courtesy HSE)

Erectors should be provided with suitable frogs on their belts to carry podgers, wrenches and other tools of their trade. Bolts and nuts should be kept in suitable containers provided at all working places or they may be kept in a pouch on the erector's belt. Tidiness and care with small items are essential; loose bolts, washers and tools should not be left on beams, walkways or working platforms. The use of toe boards on working platforms reduces the likelihood of items being kicked over the edge. Small items should never be thrown to the ground or from one erector to another.

Stacked materials. Accidents may arise from the collapse or fall of poorly stacked steelwork in the storage area. Lighter members should be stored on purpose-made racks of adequate stability. The height of stacked materials not stored on racks should be restricted and the materials should be supported on timber battens with packing and wedges to prevent the members slipping and the stack collapsing. Materials should be stacked clear of the ground because deposits of mud on them may cause a slipping hazard when they are in position at a higher level. The stacking should be arranged so that slings can be fixed and removed safely and safe means of access should be provided where it is necessary to climb on to a stack for selection and slinging.

Preventing framework collapses

At all stages during the erection phase most frameworks, or parts of them, are in unstable equilibrium. This may be due to

- the geometry of a series of pin-jointed frames

- the dependence for stability on the structural interlocking of other elements such as cross walls, floors, cladding panels or other elements to be erected at a later stage
- the temporary release of bolts and fastenings to facilitate aligning and levelling.

Relatively small forces on a structure due, for example, to the slight impact of a crane load or jib, eccentric loading from material stacked on the framework, overtightening of guy ropes, freeing jammed members, springing or pulling members into position, wind and even someone climbing a ladder positioned against a light framework, may be sufficient to cause collapse. Adherence to a properly prepared method statement will overcome these problems. There is also a need for the designer to convey to the erector, information about temporary instability.

Framework stability. The main principle for ensuring stability during erection is to create a rigid box consisting of four columns connected at their tops by beams, the frame being prevented from distorting by bracing between the columns. Single columns must first be anchored by guys or props and wedges (in the case of precast concrete columns in foundation pockets) in four directions although when the adjacent column has been erected and connected and braced to it, the two guys in that direction may be removed.

When the rigid box has been established, the rest of the framework is built from it and effectively anchored to it. However, temporary bracing, column wedging, propping and guying will still be required for the various elements of the structure as it 'grows' (Fig. 4.24). In no circumstances should sheeting rails or purlins be relied on as temporary bracing.

Positioning bolts may need to be slackened during the lining and levelling operation. To minimise the likelihood of collapse final alignment of the framework should be carried out a bay at a time as the erection proceeds (no more than two bays should be erected before final alignment is carried out) and permanent connections then made and holding-down bolts and baseplates fully grouted. During the alignment all temporary supports and bracing should be maintained. Similarly, when joints are loosened to facilitate the positioning of other members the stability of the framework must be maintained by means of temporary guys, props or bracing.

Trusses and girders. Steel roof trusses and lattice girders should be propped and/or guyed into position before the lifting gear is released. An adjacent truss or girder should then be erected and the necessary plan bracing installed to provide stability (for girders diagonal bracing may also be required between the top flange of one and the bottom adjacent beam), connecting the two before the lifting gear is released. Reliance must not be placed on purlins between trusses to provide the required stability.

A common fault in the erection and positioning of roof members which has led to many collapses is concerned with the installation of purlins. There is a tendency, when one or two trusses have been positioned, to lift and park the majority of the purlins in one place between two adjacent

HAZARDS OF CONSTRUCTION AND THEIR PREVENTION

Fig. 4.24. Use of temporary bracing in a well-erected frame (courtesy HSE)

trusses. The erectors then take them from this stack and bolt them in their final positions between the erected trusses. A large number of purlins parked in a narrow area may be sufficient to cause an overload and, especially with the relatively slender rafters of a portal frame, may cause buckling and collapse. Moreover, erectors tend to drag the purlins from the stack across the backs of the rafters to their final position, and the force required to overcome the friction is very often sufficient to cause the overturning of a truss. Either the specially designed purlin cradle referred to earlier should be used or the quantity of stacked purlins should be limited and spread between a number of trusses — furthermore, of course, they should be lifted and not dropped into position.

Precast concrete. Precast concrete columns are usually located on level packings in pockets cast in the concrete foundation. Wooden wedges cut to match the slope of the pocket sides should be inserted on all four sides before the column is released from the crane. Stability should also be provided by four push–pull props located against bolted-on timber yokes whose lower faces are chamfered to bear fully on to each prop. The foot of each prop must be firmly anchored at floor level. When two or more columns have been erected and connected by a beam or beams, temporary bracing by taut wire rope or scaffold tubes must be installed. Propping and wedging of the columns must be left in position until the foundations are fully grouted and the permanent connections between columns, beams and rafters have been made.

With precast concrete portal frames, where the first rafter is lifted in one piece and connected to the columns the lifting gear must not be released until temporary stability is provided to the rafter by means of props or guys in two directions at right angles to the span. On large spans the rafter may be in two halves, in which case both halves need temporary support before the lifting gear is released. After the erection of the adjacent rafter the permanent bracing between the two should be installed followed by the purlins. Finally, after all the permanent connections have been completed the temporary supports, bracing and associated brackets or cleats etc. may be removed.

Safe use of cranes

The main hazards associated with crane operations on construction sites may be categorised as

- the overturning of a crane or the structural failure of one of its elements
- the dropping of the suspended load or part of it
- electrocution
- trapping of people
- incorrect erection and dismantling procedures.

These hazards are considered in respect of mobile, tower and derrick cranes, which are the types most commonly used on construction sites. More detailed information may be obtained from British Standard *Code of Practice for the Safe Use of Cranes* CP 3010 (under revision).[6]

Planning lifting operations

Virtually all crane accidents are the result of operational errors rather than inadequate design or lack of quality in the manufacturing process. It is fundamental for safety, whether for one-off jobs or routine activities, that all lifting work is carefully planned. Planning should include

- selection and briefing of the lifting team, i.e. drivers, slingers, banksmen and supervisors
- organisation of the lifts: for major lifts this would need to be very detailed
- identifying the weight and the centre of gravity of the load to obtain the correct position for lifting; where either has to be calculated this must only be done by persons competent to do so
- the maximum radius to which the load will be taken
- the maximum lift height for the load
- selection of the crane(s) and lifting gear to be used, allowing an adequate working margin for load, radius and lift height, so that the crane(s) is not working at maximum capacity for long periods
- the siting of the crane(s) and the necessary ground preparation: for the siting, the path to be taken by the load and moving parts of the crane body and jib etc. throughout the lift should be considered relative to any obstruction; it should also include ease of access and suitability for erection and dismantling operations; ground prepara-

tion should include levelling and foundation requirements; the routes to be taken by mobile cranes should be assessed for their suitability; they should be clear of obstructions and not too close to the edges of excavations or embankments; the ground must be reasonably level with a bearing capacity suitable for the wheel or track loads — sleepers or mats may be required
- special considerations for any tandem or multiple lifting (see p. 82)
- erection and dismantling procedures in strict accordance with the methods, sequence and materials laid down in the manufacturer's manual
- programming operation to avoid working under suspended loads; erection of notices warning people not to pass under suspended loads.

Selection of operatives
Drivers. All crane drivers must be at least 18 years of age, medically fit and with an aptitude for judging distance. They should have been trained in crane operation, simple maintenance and hand signalling. It is advisable that they hold a certificate of competence for the particular machine they operate. The Construction Industry Training Centre holds courses and issues certificates to those successful (*see* p. 206).

Banksmen and slingers. Banksmen and slingers must also be at least 18 years of age, medically fit and with an aptitude for judging distance. They should have been trained to a suitable level of competence. They should be readily identifiable, for example they should wear a garment that is easily visible or a distinctively coloured safety helmet.

Duties of the lifting crew. Where the driver cannot see the whole area of the lifting operation, from picking up the load, throughout its full travel to where it is to be put down, a banksman must be employed to guide the driver and to control the operation of moving the load. Communications between driver and banksman must be of the highest order and are best made by radio telephone. If visual hand signals are used both driver and banksman should be clear about their meaning. They should preferably conform to those recommended in British Standard *Code of Practice* 3010.[6] The distance between the driver and the banksman should be such that hand signals are clearly visible and there is no possibility of confusion of their intention occurring. Where this is not possible, and radio telephones are not available, more than one banksman must be used. One of the duties of the banksman is to prevent people from straying underneath the carried load. One person particularly vulnerable is the slinger. When the load is ready for lifting he should position himself out of the line of lift and travel before signalling to the banksman or driver to start lifting. He should never walk or stand under the suspended load.

Main hazards and safety measures for crane operations
Overturning or structural failure. Overturning or structural failure in cranes is caused by overloading. It is essential that the safe working load

(s.w.l.) at its associated radius is never exceeded. Sometimes the manufacturer's s.w.l. figures include the weight of the lifting gear (hook, slings, spreader, beams, etc.) and therefore the weight of these items must be deducted from the s.w.l. figures to arrive at the load that may be lifted.

The s.w.l. may be determined by the stability of the crane, i.e. when the s.w.l. is exceeded by a relatively small amount, tipping will begin. However, for some cranes the s.w.l. may be limited by the structural or mechanical strength of components such as the jib or hoist ropes. A structural failure will then occur without any indication of tipping — this is often the case for mobile cranes operating with outriggers in position and for hydraulic telescoping cantilevered-jib truck cranes. It is essential therefore, that drivers and planners do not rely on loading a crane to the point where it shows early signs of tipping in order to obtain the highest possible lifting capacity. Structural failure may occur first, suddenly and without warning.

Overloading is not necessarily a load in excess of the load specified as the s.w.l. but may occur because the allowable radius for the load is being exceeded. Some examples are

- excessive load swing due to rapid deceleration of the slewing motion by severe application of the brake; similarly, snatching the load from rest and hard braking to arrest a descending load can also cause excessive swinging
- using the crane to drag a load, usually from outside the radius associated with the s.w.l. for that load; dragging may cause a considerable increase in the load being moved because of the frictional force incurred and should never be allowed
- eccentric lifting: the hoist rope must be vertical above the centre of gravity of the load otherwise the load will tend to swing
- a wind effect on loads of relatively large surface area (e.g. shutters): it is necessary to counter this by means of tail rope(s) attached to the load and manipulated by a banksman
- differential settlement of the crane supports or foundations: with mobile cranes where the load of the crane transmitted through the wheels, tracks or outrigger feet exceeds the allowable ground bearing pressure, sinking will occur; this will almost certainly cause a swing of the jib and load; sleepers, mats, plates under outrigger feet or wheels or even the provision of a properly constructed concrete hard standing may be used to ensure that the allowable bearing pressure is not exceeded; the maximum load to be distributed must be used in any calculations, i.e. the load under a wheel, track or outrigger when the jib is operating over them; care must be exercised to ensure that the crane wheels or tracks do not approach too closely to the edges of excavations or embankments
- stretch in the jib suspension ropes: with mobile and guyed derricks when the load lifted is near the maximum s.w.l. any stretch in the jib suspension rope will allow the jib head to deflect, causing the radius to become excessive; before such a lift is made, the jib should be derricked in sufficiently to compensate for this

- rebound of the jib head when the load is grounded: when the jib is near its minimum operating radius it may swing backwards when the load is grounded because of the sudden release of tension in the jib guys; this may result in structural damage to the jib by the backstays or in the crane overturning backwards; gentle and smooth grounding of the load should overcome this hazard.

Dropping the suspended load

Correct slinging of the load and its security on the hook is of primary importance. Only slings and lifting gear of the appropriate s.w.l. with the required valid test certificate (i.e. which have been thoroughly examined within the previous six months and appropriately certified) may be used. Chains must not be knotted and shackles must be fitted with proper pins. Slings should be protected around sharp corners by fitting packing pieces. Hooks should be of sufficient capacity for the heaviest load intended to be lifted.

The load should be slung so that the hoist rope is vertically above the centre of gravity of the load to prevent tilting or swinging. Where there is some doubt about the position of the centre of gravity, the load should be trial lifted just clear of the ground. If it tilts or swings it should be lowered and the slinging readjusted.

The hook should be either of the Liverpool or C-type or be provided with a safety catch to prevent the load, sling, ring or other suspension device from being displaced. The hoist rope must be of the type and size specified in the crane manual, in good mechanical condition and properly maintained. Hoist rope failure resulting from poor maintenance has been the cause of many accidents. Broken wires and corrosion are usually the first signs of a dangerous condition in a hoist rope.

The load must be carefully and smoothly manoeuvred throughout its travel and it must not be allowed to collide either with any part of the crane or with any temporary or permanent works, plant or stacked materials. The load must be maintained vertically under the hoist rope and it may be necessary for the banksman to control the load by a tail rope to prevent swinging.

Danger of electrocution

If the jib touches or closely approaches a live overhead power line the electric charge will pass into the crane body via the jib and everyone who is in the vicinity, especially the banksman if he is steadying the load by means of a tail rope, is in danger of being electrocuted. All overhead lines must be treated as live unless the electricity authority has declared them dead.

When work is to be carried out in the vicinity of overhead lines the crane should be prohibited from approaching nearer than 6 m plus the total length of the jib and including any intended extensions measured horizontally along the ground from vertically below the nearest conductor. Allowance should be made for the lateral movement of the conductors because of the effect of wind — which is very likely to take place with overhead

Fig. 4.25. Goal posts to indicate maximum height for safe passage of cranes and other plant under high tension cables (courtesy HSE)

lines of fairly long spans. The limits should be marked on the ground by warning markers parallel to the conductors.

When cranes are to pass under overhead power lines, the allowed crossing route must be clearly marked and goal posts (Fig. 4.25) should be erected at each side of the crossing approach. They should be positioned no closer to the outermost overhead conductor on each side than the horizontal distance markers. The cross-bar should be at a safe height so that the jib must be lowered below it before making the crossing. Devices are available which can be fitted to crane jibs to give a warning when they approach within a predetermined distance of an overhead conductor. However, they have limitations, and their availability should not preclude the use of horizontal markers and goal posts. When work is to be carried out within 15 m of overhead lines suspended from steel towers or 9 m from those suspended from concrete, timber or steel poles, the Electricity Board should be advised well in advance.

Clear instruction on the actions to be taken if the crane becomes live must be given to every crane driver and a notice giving the information should be prominently displayed in his cab. These instructions should inform the driver what to do if the crane makes contact or becomes live. The driver should

- remain in his cab
- warn others to keep away and not to touch any part of the crane, rope or load
- ask someone to inform the electricity authority
- try to move the crane away from the power line

- where this is not possible, stay in the cab until safe conditions are confirmed
- where it is essential that the driver leave his cab, he should jump clear and not climb down; he should not touch any part of the crane or load and the ground at the same time.

Trapping

During slewing operations, many serious accidents have occurred with people being trapped between the moving superstructure of a mobile crane and any nearby fixed object such as adjacent structures, plant or stored materials. A clear passageway of 600 m must be maintained between any part of the crane liable to move and any fixed object, otherwise the gap must be effectively fenced off.

Slingers are the likely victims of the other main trapping hazard, with injuries to fingers, hands and feet when they become trapped between the hook and the sling or other lifting gear as the crane takes the load. The driver must not operate any crane motion until the slinger has given the appropriate signal from a safe place. Moreover, the slinger must not grasp the hoist rope (e.g. to guide the load) at the instant of lift or just before grounding since trapping between the rope and hook block sheave is likely. A pole or tail rope should preferably be used to guide the load but if the hand is used it should be the flat of the hand against the load. When guiding the load, just before final grounding the slinger must ensure that his feet are well away from the load.

Erection and dismantling

Many accidents happen during the erection and dismantling of cranes, including erectors falling, being struck by falling tools or materials, and being trapped. Only erectors trained and experienced in the particular make and model of crane should be used for this work and they must be under the immediate supervision of a competent supervisor experienced in all aspects of the operation. The work must be carried out strictly in accordance with the manufacturer's procedures manual. Where the jibs can be extended by the introduction of intermediate sections, the jib should be properly supported on each side of the joint being disconnected before any connecting pins are knocked out, to prevent the jib falling. Also, the inserted sections must be put in the correct way round and in the correct place in the jib assembly.

All erectors should wear safety helmets and, especially in the case of tower and derrick cranes, safety harnesses and lanyards should be worn and used when working at height (Fig. 4.26). With tower cranes, at each stage of the erection, the necessary counterweights or kentledge must be secured in the appropriate position. Where the counterbalance is made up of individual weights on the counter jib, these must be bolted together or otherwise prevented from becoming displaced. All pieces of kentledge and counterbalance should have their weight clearly marked on them. Before any section is lifted from the horizontal to the vertical, a check should be made for any loose tools or materials.

Fig. 4.26. Safety harness and lanyard used when working on the jib of a tower crane (courtesy Latchways Ltd)

Safety measures for specific types of crane

Some additional points on overturning and structural failure hazards and safety measures associated with specific types of crane are now dealt with.

Mobile cranes

Operating on a slope. When mobile cranes are travelling, or working in a static position on a gradient, where the load is on the downhill side the operating radius is increased. Where static work is carried out across a slope the load will swing sideways. This circumstance should be avoided. Static work should only be carried out on level ground or when the crane has been properly levelled.

The slewing ring of the crane must always be level before any lifting operations are carried out. This must be checked at regular intervals.

Where there is a need to travel on a gradient either the load should be carried on the uphill side (for travel either up or down the slope) or, where the load has to be carried on the downhill side, the operating radius should be reduced to compensate for the slope, and in both circumstances the slewing brake should be engaged. In either case the driving should be slow and steady commensurate with the ground surface.

Operating with outriggers not full extended. Where outriggers are fitted, where they are not fully extended (they should be pinned in position), the s.w.l. associated with the extended outriggers is not available and, if it is used, an overload is inevitable. Outriggers must always be fully extended and jacked down on to suitable plates and packings in order to distribute the load (Fig. 4.27).

Operating with under-inflated tyres. Where the tyres of a wheeled mobile crane are not properly and evenly inflated the crane will not be operating on a level base and an excessive radius will occur.

Assessing the strength of jib and suspension ropes in piling operations. A particularly onerous duty for mobile cranes on construction sites is their use in piling operations. The crane must be of adequate capacity safely to accommodate the weight of the piling equipment (e.g. hammer, extractor and leaders). When the crane is moving from one location to the next with leaders or frame suspended from the jib, the suspended equipment must be prevented from swinging and subjecting the jib to undue lateral and dynamic (bounce) loading.

When sheet piles or the casings of bored piles are being extracted an extractor should be used to overcome the friction between the ground and the pile or casing. Final extraction should be carried out by a smooth steady pull on the hoist rope. This is a very heavy duty which should be restricted to machines that are designed for it. When the required capacity of the crane is being assessed for extracting sheet piles, the friction between adjacent piles should be allowed for. It is of paramount importance with this type of work that the pull is exerted on a vertical hoist rope.

Fig. 4.27. Mobile crane with outriggers fully extended and jacked down on a firm base (courtesy HSE)

94 CONSTRUCTION SAFETY HANDBOOK

Fig. 4.28. Audio-visual warning devices were fitted to the tower cranes at the Thorp project at Sellafield to warn drivers of possible collision of jibs, ropes, counterweights or loads (courtesy Balfour Beatty Construction Ltd)

Tower cranes

Wind effects. Tower cranes are specially vulnerable to the effects of wind and must not be operated outside the limits stipulated by the manufacturer. An anemometer should be fitted at the highest suitable position with its indicator in the cab clearly visible to the driver. When lifting operations are suspended for any reason, all loads should be taken off the hook and the hook should be raised to its highest position. The main jib should be positioned away from the wind and put into free slew. The power supplies to all motions should be switched off and the hoisting and trolleying (but not the slewing) brakes or locks applied. In adverse wind conditions the wheels of rail-mounted cranes should be clamped or movement prevented by other means. No advertisement boards should be attached to any part of a tower crane without the manufacturer's approval and then only in strict accordance with the restrictions imposed.

Collision between tower cranes. When more than one tower crane is to be used on a site, where they are in close proximity to each other there is a danger of jib collision, jib–hoist rope collision and counter jib–hoist rope collision. Careful planning and skillful and well-controlled crane operations are necessary if serious accidents from overturning, structural failure or load displacement are to be avoided (Fig. 4.28).

Where possible, cranes should be sited so that jibs, hoist ropes and counter jibs cannot touch. Where this is not possible, even when the jibs are at different heights, careful control of all lifting operations is necessary with strict adherence to the programmed lifting sequences. These must be agreed at the planning stages and controlled by one person. Good direct communication between all drivers and the ground controller is essential and the maintenance of communications equipment is vital.

Devices are available to give warning when the crane is approaching an obstruction but these should only be used as additional safeguards. Where more than one rail-mounted crane is operated on the same track, cut-out switches should be installed on the track to prevent collision.

Foundations for tower cranes. Foundations for tower cranes must be properly designed to CP 3010 and in accordance with the manufacturer's recommendations. It is particularly important that the allowable bearing pressure of the ground is not exceeded and that the bases are well drained and maintained during operations.

For rail-mounted cranes, where the rails are supported by sleepers, these must be spaced to support the maximum crane wheel loading and to resist any tendency to heel or turn the sleepers over. Sleepers should be at right angles to the tracks and should extend sufficiently at both sides of each rail to ensure that the load is properly spread to the foundation. Where a layer of hardcore and ballast is used beneath sleepers, it should be maintained to provide a solid and compact support. It should not be packed high in the centre of the sleepers so as to cause convex bending.

The tracks should be laid straight or to radii of the correct curve to suit the types of bogie, and they should be tightly butted end to end. They should be in good condition. The gauge should be maintained by means of tie rods which should be able to withstand the compressive as well as the tensile forces. This can be done by threading the tie rod through scaffold tubes which have been cut to the correct length between the inside webs of the rails.

Fish plates should be fitted to both sides of the rails and secured by bolts. The holes in the rails to accommodate the fish plates should be drilled, not burned, and located over sleepers. Security from running off the end of the track should be provided first by travel limit switches placed near the end of the track, then by sandboxes to reduce the travelling inertia of the crane if it overruns the limit switches. Finally, at a distance equal to half the crane base length from each end, crane stops should be securely positioned.

Climbing cranes are often supported by the permanent structure and it is important that the erectors check with the client or their professional advisers to ensure that the structure is capable of withstanding the expected loads.

Danger of electrocution. Where the crane is electrically operated from a source external to the crane, the structure and all electrical equipment must be properly earthed. For rail-mounted cranes at least one of the rails must be electrically bonded across every joint and the rail itself properly earthed.

Power supply cables should be of armoured construction or otherwise protected from mechanical damage (e.g. run in conduit), and where the protection is provided by a conducting material it should be bonded to earth at each end.

Access for personnel. Good access to reduce the possibility of falling should be provided for all those who need to climb the crane (e.g. drivers, maintenance workers and inspectors). A permanent steel ladder is required, securely fixed and with safety hoops, and extending from the ground to the driving cab. Wherever possible the ladder should be within the mast, and must include landings at 9·14 m intervals for people to rest safely.

Access through the slewing ring presents a trapping hazard between the fixed and slewing parts of the mast. Devices should be fitted to enable the slewing brake to be controlled from below the trapping area. Alternatively, access may be provided from just below the slewing ring by means of a ladder outside the mast. This presents a falling hazard when people are climbing from the ladder inside to the access platform outside the mast, and strong wire mesh barriers between guard rails and toe boards should be provided on the platform.

Safe means of access is necessary along the jib and counter jib. Any gangway or platform at cathead level, together with walkways (usually open mesh) along the jibs, should be provided with guard rails and toe boards complying with the Construction (Working Places) Regulations. Alternative access may be provided by a lightweight cage that is suspended from the jibs and can be traversed along them. In such a case it is safer if movement control is operated from the cage itself. However, it will then be necessary to incorporate interlocks so that the saddle traverse control in the cab is locked off when the jib cage is in use.

Where neither method is practicable everyone on the jibs must wear safety harnesses fitted with two lanyards, so that when negotiating obstacles the wearer can be secured at all times either to the structure or to a specially fixed taut steel-wire rope of adequate strength to withstand the shock of a man falling.

Scotch derricks

Incorrect position of the jib. The angle between the backstays of a scotch derrick is approximately 90° and no attempt must be made to erect the jib and to lift a load within this quadrant. This action might result in buckling of the backstays, uplift on the kingpost and overturning.

Kentledge. When the scotch derrick is rail mounted on bogies, the correct amount of kentledge should be secured to the backstay bogies to prevent tipping under maximum load, and the total weight of the kentledge should be clearly marked on each bogie. Also, each individual block of kentledge should be marked with its weight.

Bracing of gabbards. Additionally, when the scotch derrick is mounted on gabbards, either fixed or rail mounted on bogies, the gabbards must be adequately braced within themselves and between each other (except across the bowsill side) to overcome buckling or twisting.

Derricking under control. For a single-motor scotch derrick an interlock must be fitted between the derricking clutch and the pawl sustaining the derricking drum so that neither can be disengaged unless the other is fully engaged, otherwise the jib with its load can fall.

HAZARDS OF CONSTRUCTION AND THEIR PREVENTION

Guyed derricks

Some additional points concerning overturning and structural failure hazards and appropriate safety measures with guyed derricks include the following.

Unless the guy ropes are of the correct size, and are correctly spaced and properly anchored, the mast may topple. The security of the guy ropes should be ensured by means of anchor stirrups or plates set in a concrete foundation. The guy ropes should be equally spaced radially around the mast and a minimum of five should be used. The vertical angle between the mast and guy should be not less than 45°. The ropes, which are usually fitted with turnbuckles, must not be over-tensioned — all tensioning should be carried out by a competent person experienced in undertaking the functions.

A chart showing safe working loads at different radii for various jib lengths should be attached to the mast. This should also indicate the rope sizes, their angle of inclination and the anchor weights for each guy.

Testing and examination of cranes

In Britain it is a statutory requirement that all cranes on construction sites shall

- be tested and thoroughly examined before being taken into use, every four years and before use after any substantial alteration and repair
- be inspected weekly
- be thoroughly examined every 14 months
- have the anchorage or ballasting arrangements
  - examined on each occasion before erection
  - examined after adverse weather conditions that are likely to have affected their stability
  - tested after each erection on site and after any adjustments that involve changes to the ballasting or anchorage arrangements or their security
- when an automatic safe load indicator is required, have it
  - tested before the crane is taken into use and on each occasion after it has been wholly or partially dismantled
  - tested after each erection or any alteration of the crane likely to have affected the proper operation of the indicator
  - inspected weekly.

All these tests, inspections and examinations must be carried out by competent and experienced people, and the results must be entered in the appropriate register for the first three requirements and on Form 91 for the last two.

Other safety measures

Automatic safe load indicators. Automatic safe load indicators are safety devices which provide a visual warning to the driver when the crane is lifting between 90% and $97\frac{1}{2}$% of its maximum safe working load. An audible warning is given to the driver and is sufficiently loud to be heard by those in the vicinity of the crane when the crane is lifting between $102\frac{1}{2}\%$ and 110% of the safe working load, i.e. when it is overloaded.

In Britain it is a statutory requirement to fit these devices to all cranes used on construction sites which have a lifting capacity over 1 t. It is very important that they should be maintained in good working order and that they should never be immobilised or tampered with so that they do not function correctly when the crane is in use.

Load radius indicators. Load radius indicators are indicators which show the radius of the jib, trolley or crab of a crane at all times and the safe working load corresponding to that radius. It is a statutory requirement in Britain to fit them to all cranes which have a variable operating radius.

Transport and mobile plant

Of all the construction industry fatalities for 1981–85, 21% involved transport and mobile plant. The deaths included those for loading and unloading operations and people falling from plant and vehicles.[7] The types of vehicle involved in the accidents included excavators and shovels, earthmoving equipment (i.e. crawler tractors and bulldozers, scrapers and graders), dumpers and dump trucks, lorries, vans, cars, buses, motor cycles, mobile cranes, rough terrain forklift trucks, locomotives and rail waggons, road rollers and mobile working platforms.

Well over half of site transport accidents involve people being run over, struck or crushed by vehicles moving forward or reversing, collisions between vehicles and collisions of vehicles with fixed objects (such as falsework and scaffolding). This category includes those accidents where vehicles and plant overturn by falling into excavations and down embankments. Other accidents occur when people fall from vehicles not designed to carry passengers and when equipment or material falls from vehicles when they are moving or stationary.

The hazards of working with transport and mobile plant on site are numerous but the fundamental cause of most accidents is the lack of adequate planning and the failure of management to organise safe systems of working and to instruct, train and properly supervise personnel.

Site transport planning

Hazards with site transport and mobile plant can be reduced by

- the segregation of vehicles and pedestrians by barriers and the segregation of vehicles from fixed objects, including scaffolding and falsework, by the use of barriers and/or fenders and effective signing; heavy earthmoving equipment should also be segregated from lorries and light traffic whenever possible
- the provision of clearly signed and indicated routes around and into the site and safe parking areas; site roads and haul routes should be of adequate width to allow approaching vehicles to pass safely; sharp bends should be avoided or the road should be locally

Fig. 4.29. Wheel stops for vehicles tipping into excavations (courtesy George Wimpey and Co. Ltd and CITB)

widened; haul roads crossing public roads should be controlled by traffic lights or manually operated stop–go signs
- effective procedures for dealing with reversing, e.g.
  - use of banksmen
  - provision of adequate turning spaces
  - organising one-way systems for loading and unloading to eliminate the need for reversing
  - fitting audible warnings when reverse gear is engaged
  - excluding pedestrians, except the banksmen, from areas where vehicles have to reverse
  - fitting audible proximity warnings or closed-circuit television in vehicle cabs
  - providing reversing stops for vehicles tipping into excavations (Fig. 4.29)
- good effective lighting of traffic routes and work areas and, where there are obstructions along the route such as scaffolding and falsework, plant buildings and materials
- regular maintenance of site road surfaces by grading, repair of pot holes etc. and dust control by regular spraying with water
- preventing vehicles from approaching close to the edges of excavations by effective barriers and good road marking; where vehicles are required to approach excavations or embankment edges to deliver materials, for example, the reversing and tipping operation should be controlled by a banksman and effective wheel stops should be used

- the erection of goal-post height gauges at approaches to overhead obstructions, e.g. electricity lines
- clear indication of the routes of underground cables, wires and pipes

Other measures that must be taken by management to provide for the safety of personnel include:

- the provision of a scheme for assessing the competence of drivers and ensuring that only those employees who have been authorised to do so are allowed to drive vehicles or to operate plant; nobody under the age of 18 may drive vehicles on site
- the imposition and strict enforcement of speed limits and good driving behaviour
- the provision of loading controls for vehicles to ensure they are not incorrectly or overloaded; the use of loading height gauges where appropriate
- imposing a ruling that no loading or unloading of lorries takes place while the driver is in his cab and the banksman and/or plant operator prevents people from straying under any load being slewed during loading or unloading
- the use of construction plant fitted with roll-over protection structures (ROPS) for operations where roll-over is a possibility (*see* ROPS below)
- the use of vehicles and plant fitted with falling object protection structures (FOPS) wherever drivers may be struck by falling material, e.g. on demolition sites (*see* FOPS below)
- training for all site employees covering the hazards and precautions of working with site transport, followed by effective supervision
- the provision and use of high visibility garments for all those working in the vicinity of site transport, especially banksmen
- the provision of a planned inspection and maintenance scheme for all vehicles and plant, with properly kept records
- ensuring that all subcontractors, suppliers and the self-employed meet the same standards of training and supervision as contractors' employees and that they comply with all the rules laid down in relation to the control of transport on site.

The special hazards and precautions to be taken with rail transport on site are described under the section on tunnelling.

ROPS and FOPS. To implement two EC Directives the following regulations for roll-over protection structures (ROPS) and falling object protection structures (FOPS) for construction plant are applicable in Great Britain.

- The Roll-over Protective Structures for Construction Plant (EC Requirements) Regulations 1988, SI 1988, No. 363. The regulations prohibit the marketing of construction plant covered by ISO Standard 3471/1 1986 (identical to BS 5527: 1987[8]) manufactured after May 1990 unless it is fitted with ROPS which complies with the Regulations.

Fig. 4.30. Segregation of work activities from road traffic during M25 widening in Surrey (courtesy Balfour Beatty Construction Ltd)

- The Falling Object Protective Structures for Construction Plant (EC Requirements) Regulations 1988, SI 1988, No. 362. The regulations prohibit the marketing of plant covered by ISO Standard 3449: 1984 (identical to BS 5526: 1985[9]) manufactured after June 1990 unless it is fitted with FOPS complying with the Regulations.

Market means sale, lease, hire or hire purchase. The regulations require that for each ROPS or FOPS there must be an EC-type examination certificate and a certificate of conformity which must be displayed on the equipment. Those convicted of contravening the regulations, whether a body corporate or a senior person acting for it, or both, may be fined up to £2000.

Although ISO 3471/1 1986 includes off-highway dump trucks, it is not intended that the regulations for fitting ROPS (or FOPS) apply to small-site dumpers. Even though the regulations only require facilities on the plant for the fitting of FOPS, when the plant is used in places where falling objects or material may be expected, e.g. demolition sites, FOPS must actually be fitted.

Roadworks

An aspect of civil engineering closely affecting, and affected by, members of the public is highway work. Many users of the public highway are killed or injured in traffic accidents at roadworks (see Fig. 6.4) and the safety of roadworks personnel is also of great concern. The number of accidents may be reduced by adequate traffic signing (Fig. 4.30) and by rigid adherence to a number of simple rules and recommendations.

Safety of personnel

Notes for guidance in relation to the implementation of the requirements of HSW Act so far as they affect personnel who are required to undertake work on motorways and trunk roads are published jointly by the County Surveyor's Society and the Department of Transport (DTp).[10] The general safety notes are applicable to all personnel and they can also apply to work on roads other than motorways and trunk roads. Abbreviated, they state that

- all personnel must wear high visibility garments with retro-reflective markings
- inexperienced personnel must be accompanied by experienced, trained operatives
- only trained operatives may set out signs and cones
- signing must be in accordance with the *Traffic signs manual*[11] (chapter 8, Traffic safety measures at roadworks)
- prescribed signs must be illuminated; non-prescribed signs must be illuminated or reflectorised as required
- the minimum clearance between the working area and the trafficked carriageway required by the *Traffic signs manual*[11] and the DTp must be observed (see Fig. 6.4)
- the work must be undertaken in such a way that hazards to personnel and the public are minimised
- plant and materials must not intrude into any area reserved for pedestrians, cyclists or other traffic; the problems of the disabled and visually handicapped must be considered
- personnel must only cross a trafficked carriageway when it is essential to do so
- when coming from a moving vehicle, personnel should take care not to intrude into a trafficked lane
- unless it is unavoidable, signs should not be off-loaded into a trafficked lane.

Other recommendations include

- reversing of vehicles should generally be avoided unless under the guidance of a banksman
- headlamps should not be used when the vehicle is facing oncoming traffic
- rules relating to parking and manoeuvring of vehicles at roadworks must be observed
- distinctive amber lamps (rotating amber flashers) must be switched on
    o where vehicles have to stand

- o before manoeuvring into or out of traffic
- o during short-term low-speed mobile operations
- a regulation highway maintenance sign must be carried
- vehicle and plant colour together with visibility requirements should be met.

All staff and operatives engaged on roadworks should wear safety helmets not only for head protection but also because their conspicuous colour makes the wearer more readily visible to vehicle drivers and plant operators.

Traffic safety at roadworks

Traffic safety measures at roadworks are covered by chapter 8 of the *Traffic signs manual*[11] prepared by the Department of Transport, the Scottish Development Department and the Welsh Office and published by HMSO.

Authorities, contractors and others have a civil law liability to give warnings of obstructions on the highway caused in connection with roadworks. They also have to remove redundant signs. Notwithstanding any civil law liability, failure to erect signs etc. could constitute an offence under the Highways Act 1959 and possibly also under the HSW Act 1974. The type and layout of traffic signs and road markings for roadworks are described in section 4 of chapter 8 of the *Traffic signs manual*. If it is necessary to hold down temporary signs, sandbags should be used. The practice of holding down temporary signs with kerbstones or other heavy objects can be very dangerous if a collision occurs.

The boundaries of all roadworks must be clearly delineated to indicate to drivers the limits of the carriageway and to protect workmen and the works. This should be done by traffic cones and barriers and, after dark, lamps must be used as well. This is described in section 5 of chapter 8 of the *Traffic signs manual*. Barriers to protect pedestrians on the highway must be fixed, secure and immovable so that, for example, a visually disabled person is prevented from passing through.

Traffic signs and safety measures for minor works on minor roads are described in the Department of Transport's advice note TA 6/80.[12]

Underground services

The proliferation of underground cables, pipes and ducts on highway works is a particular concern. Contract documents usually include the special requirements of any public utilities present, describing the hazards and explaining to contractors the notes for working near live apparatus (*see* p. 163). Reference should be made to HSE guidance note GS 33, *Avoiding danger from buried electricity cables* and *Precautions to be taken when carrying out work in the vicinity of underground gas pipes* (British Gas Corporation).

Apart from the dangers with electricity cables and gas pipes there is the risk of flooding where water mains are severed, and of pollution where sewers are damaged.

The first essential measure towards making the site safe when excavations have to be carried out is to contact British Telecom, the gas, water and electricity boards and the local authority to advise them of the work to be undertaken and to obtain as much information as possible. Contact should be made at least one month in advance of any site work. Thereafter, it is necessary to keep in close liaison with their designated representatives. When location of the services is permitted by the authority concerned, the initial excavation must be carried out by hand and with great care.

Tunnelling
A high standard of good housekeeping (i.e. site tidiness) and site discipline (i.e. the observance of laid down operational safety procedures) is essential to eliminate or minimise the hazards implicit in all types of tunnelling operation. Compliance with procedures is particularly important in tunnelling because of the nature of the work, which is carried out in a confined area. Before tunnel construction starts a hazard analysis study should be made and the working procedures planned to overcome as far as possible the problems that are foreseen. The work plan should include both evacuation procedures and rescue operations for accidents and emergencies, e.g. a fire. Contingency planning should include discussions with the local fire brigade and ambulance service, and arrangements should be made for them to visit the site at regular intervals to familiarise themselves with access to the site and tunnel face. It is also useful to rehearse the proposed rescue arrangements for likely emergencies. It is a very skilled job and must only be carried out with the necessary expertise.

Where tunnels are being driven in air at pressures above atmospheric pressure (compressed air working) the fire brigade or ambulance personnel or both may not agree to enter the workings even in an emergency. It is therefore necessary for the contractor to provide a trained fire-fighting and rescue team, properly equipped and readily available at all times while anyone is in the tunnel.

Hazards of tunnelling

Burial from ground collapse. The general principle for avoiding ground failure is to keep the area unsupported to a minimum consistent with its type and self-supporting quality. In headings and small hand-driven tunnels up to about 3·5 m in diameter the roof, sides and face may be supported by conventional timbering, as described in BS 6164 *Safety in tunnelling in the construction industry.*[13] This must be carried out only by those with the necessary experience and expertise.

For headings timbering provides the support throughout their life (i.e. until they are backfilled) but for small tunnels a permanent lining of cast iron or reinforced concrete rings is erected immediately behind the face within the safety of the timbering. The face boards are removed as required, with excavation carried out progressively downwards from the top. In poor ground the headboards are driven forward from the previous frame to provide roof support before any excavation. The timber must be of good quality and examined before use to ensure that it is sound and free from faults.

For small headings care is necessary to ensure that completed timbering is not damaged by, for example, skips or barrows used to convey spoil from the face. Tracks should be laid and maintained in line to allow sufficient clearance between the bogies and the timbering. The timber support itself must also be inspected at least once a day and thoroughly examined at least once in seven days to be sure that it is performing its function and has not suffered damage.

Larger tunnels are constructed with shields to provide the initial protection from ground failure in the roof and at the sides. The shields comprise a stiffened steel cylinder incorporating a front cutting edge. Partial excavation of the upper part of the face is carried out after removal of the necessary face boards — the remaining boards protect the operatives from failure of the face. The shield is then pushed forward by means of hydraulic jacks approximately the width of a lining segment. The shield then protects the operatives from roof failure, and the remainder of the face is excavated in safety. After the excavation phase, the face must immediately be resupported by timbering and at the same time the next ring of segments can be erected within the protection of the back part of the shield.

In better ground the face may not need timbering and excavation may be carried out with mechanical excavators or boom-mounted cutters (roadheaders) but again advancing only one ring width at a time. Full-face tunnel-boring machines incorporating a revolving cutter head within the shield may be used in soft rock. The cutters which effect the excavation are mounted on radial arms and the spoil is directed by scoops on to a conveyor belt which carries it through to the rear of the machine. The segmental lining is again erected within the safety of the shield. Hard rock tunnels may be excavated by boom-mounted cutters, by full-face tunnel-boring machines or by means of explosives. Where support is required it may be provided by the use of steel arch ribs at intervals supporting timber or steel poling boards wedged hard to the rock face.

When explosives are used loose rock may be left in the roof or sides which could fall and cause serious injury. Therefore as soon as the area has been ventilated sufficiently to enable operatives to enter, the exposed roof and sides require careful examination, and where necessary loose rock pared down before the support work is carried out. Other methods of making doubtful areas of hard rock tunnels safe are by means of rock bolts or by sprayed concrete.

Inundation, and collapse of the tunnel face. In tunnel construction with conventional timbering whenever even a trickle of water appears at the face it should be sealed. With full-face tunnel-boring machines the construction of the machine provides face support for most of the area. There is, however, a space in the front through which the conveyor gathers the spoil. Facilities need to be provided such that at the press of a panic button the conveyor is instantly retracted and a door closes the aperture.

Where inundation is a possibility it is vital that a properly planned escape route is maintained to allow all in the tunnel to be evacuated. Well-lit uncluttered walkways should be provided at the highest level practicable. Emergency lighting should also be provided.

In suspect ground, probing ahead should be undertaken to provide information about the need for ground treatment to prevent inundation. The equipment should include suitable non-return valves to prevent the possibility of a water inrush.

Falls from platforms and the danger from falling material. Hand excavation, timbering of the face, lining erection and other subsidiary work may be carried out from working platforms. The platforms should be properly designed and provided with close-boarded deckings, guard and rails and toe boards to prevent operatives and materials falling from the working position. Scaffolding material is unlikely to be suitable for the construction of platforms near the working face as it is not sufficiently robust to withstand the continuous knocks from falling material.

Atmospheric pollution. The types of dangerous atmosphere that may be encountered are those due to

- oxygen deficiency (*see* pp. 49, 112, 113, 116 and 134)
- carbon monoxide fumes, the usual source of which (other than fires) is the exhaust fumes of plant driven by petrol or diesel engines; petrol engines should never be used in tunnels during construction; compressors and any other static plant at ground level should be sited so that their exhaust fumes do not enter any access shaft, ventilation shaft or adit connected to the tunnel workings — it may be necessary to fit an extension pipe to the exhaust to convey the fumes away and down wind
- carbon dioxide, which may occur naturally, particularly in limestone areas, and is associated with oxygen deficiency
- methane, which occurs naturally and must be expected in coal measures, peat areas, carbonaceous strata and in or near refuse dumps (*see also* p. 50)
- hydrogen sulphide, which occurs naturally as a result of the decay of organic matter containing sulphur
- nitrous fumes, produced as a result of blasting operations and from welding.

Another atmospheric pollutant common in tunnelling is the respirable dust created by the excavation process and subsequent operations, dust which may be dangerous and may cause permanent damage to the lungs; moreover, the likelihood of bronchitis may be increased for those exposed to excessive concentrations of dust. The occupational exposure limits for all substances are given in guidance note EH40 published in Britain by the Health and Safety Executive. The types of dusts most likely to be encountered are

- silica (e.g. from sandstone, granite, flint and fossils), which is dangerous and may cause silicosis; another source may be spray concreting operations

HAZARDS OF CONSTRUCTION AND THEIR PREVENTION   107

Fig. 4.31. Good ventilation on land drive through Castle Hill where NATM was used provides a clear atmosphere in the Channel Tunnel workings (courtesy TML and Eurotunnel)

- coal dust (from coal measures), which is dangerous and may lead to pneumoconiosis
- general dust (e.g. chalk, dry clay or cement), excessive exposure to which may lead to bronchitis; furthermore, fossils and flints (which are siliceous) may be found in chalk, for example.

The danger from atmospheric pollution may be averted or considerably reduced by the provision of an adequate system of ventilation (Fig. 4.31).

Where dust is created by the excavation process, particularly where boom cutters are used, suppression by water spray may solve or substantially reduce the problem. Otherwise exhaust ventilation close to the source is required. Where the concentrations cannot be reduced below the recommended exposure limits appropriate protective breathing equipment must be worn. This may apply more often to those operations not directly concerned with the excavation process, e.g. spray concreting. Frequent routine atmospheric sampling and testing for the presence of dangerous levels of gases and dust likely to be encountered must be made.

When there has been a break in operations and the ventilation system has been shut down for several hours, for example over a weekend, a holiday or even after a breakdown in the ventilation system for a few hours, the atmosphere over the full extent of the workings must be tested by a competent person wearing breathing apparatus. The atmosphere must be declared safe before any other person is allowed to enter.

Moreover, where particular operations, not necessarily continuous, such as spray concreting (shotcreting) and blasting are undertaken, specific atmospheric checks should be made. In the case of blasting, the checks should be made a few minutes after shot firing, and operatives should not be allowed back to the face to clear rubble etc. until the atmosphere has been declared safe. In all cases the necessary action to be taken will depend on the results of the tests and may require the evacuation of the workings until conditions are declared safe.

Further precautions include the following.

- There is a need to maintain effective communication between ground level and all working points in the tunnel. In relatively shallow shafts and short headings direct voice contact may be satisfactory or a simple sound signalling system by bell or whistle. In the larger and longer tunnels obviously more sophisticated systems are necessary. Frequent checks should be made from the surface to ensure that people in the workings are well and that the equipment is functioning correctly. This is particularly appropriate with small tunnels and headings where the majority of the work is at the face and carried out by a few operatives.
- Appropriate rescue and first aid equipment including suitable breathing apparatus should be kept at ground level ready for immediate use near the access shaft.
- In the event of someone in the workings being overcome by a dangerous atmosphere entry for rescue must only be made by people wearing suitable breathing apparatus. In no circumstances should any rescue attempt be made by people not protected in this way.

Trapping. Where belt conveyors are used for muck disposal all dangerous parts such as chains, sprockets, gears and in-running nips — particularly between the belt and the head and tail pulleys — must be enclosed, or otherwise securely guarded. A continuous emergency stop switch cord should be provided along the whole length of the conveyor and be easily accessible. No maintenance work should be carried out on a conveyor in motion and an isolator switch should be provided for use when maintenance work is undertaken. There should be only one position from which the conveyor could be started. A distinct audible signal, understood by all who enter the workings, should be given, followed by a time interval to allow people to get clear, before every start-up. Riding on the conveyor belt should be prohibited.

The final positioning of all tunnel lining segments usually involves manual handling, and a great deal of care is necessary to avoid hand (or fingers) becoming trapped between units or between a unit and fixed equipment. Suitable hand tools for insertion in guidance holes will minimise this risk. In large tunnels where the segments are very heavy, special segment-erecting machines are employed but even with these the final positioning often requires manual assistance.

Transport. Where muck, material and men are conveyed by rail transport on narrow-gauge track the main dangers are that people may be trapped and crushed between the waggons (or loco) and the sides of the tunnel or between waggon and waggon in the case of a twin-track system, collision with other plant or equipment in the tunnel and derailment.

Unless the size of the tunnel allows adequate clearance between the sides of the tunnel and the widest waggon (or loco), allowing for the swaying motion of the waggons, refuges should be provided at 18 m intervals along the tunnel to shelter people in safety while the train passes. They may be cut into the side of the tunnel or be prefabricated platforms fixed to the tunnel sides with guard rails and easy access. British Standard 6164[12] quotes 500 mm as the desirable minimum clearance between any part of a vehicle and any fixed equipment (that may be attached to the side of the tunnel). For the loco driver it is also important that there is adequate overhead clearance, which should be not less than 1·1 m from the seat of the loco.

Clear audible and visual warning of approach from the loco will enable people in the tunnel to take refuge and of course an efficient train braking system must be provided and maintained. The track should be of adequate section, properly supported, laid to line, level and gauge. It should have an even running surface and the rails should be properly jointed with bolted fishplates or other means. Poor track construction and maintenance together with excessive speed, especially on curves, are possibly the principal causes of derailment, and often the cause of collision with a fixture at the side of the track.

In small headings where muck skips may be conveyed on narrow-gauge track the maintenance of good clearances between the skips and the side is important for the safety of the timbering. A particular concern is the possible collision with a gas cylinder that may be for use in the tunnel. Storage of gas cylinders underground should be avoided and they should only be taken down for immediate use; even then they should be properly stored in a protected container well clear of any rail track.

It is vitally important for the loco driver to be trained and experienced and for him to be able to see ahead of the train. Thus both good lighting in the tunnel and a headlight on the train should be provided. Where the driver cannot see ahead of the train because he is pushing larger waggons in front of his loco, a banksman, in communication with the driver, should proceed on foot in front of the train to ensure clearance for the train to move forward.

In the narrow confines of tunnels great care must be exercised by non-rail transport drivers (e.g. dumpers) to avoid colliding with other vehicles and pedestrians. Again, good lighting and experienced drivers are essential for the safety of operations. The vehicles themselves should be fitted with good head and rear lights which are visible at 60 m, together with audible and preferably also visual warning devices indicating that the vehicle is about to move and a distinctive audible warning for reversing operations.

Noise. Noise should be kept to a minimum by the machinery design and by good maintenance. Where the noise levels cannot be reduced to an acceptable level the use of ear protectors should be used. Pneumatic tools, often employed in hand drives, may be substantially quietened by the use of muffle bags.

Electrical hazards. Electrocution and electrically initiated fires may be prevented by the proper design, installation and maintenance of all electrical equipment. Comprehensive guidance is provided by BS 6164.[13] In small headings where perhaps the only electrical equipment is the lighting, this is usually of the temporary festoon type. It should be supplied at a voltage not exceeding 50 V a.c., with a maximum voltage to earth not exceeding 25 V a.c. Where for any installation there is a possibility of methane gas entering the workings only explosion-protected electrical equipment must be used.

Compressed air hazards. The main hazards associated with tunnels driven in compressed air are

- decompression sickness, including aseptic bone necrosis
- increased risk of fire because of the greater concentration of oxygen
- inundation.

The likelihood of decompression sickness can be minimised by strict adherence to correct compression and decompression procedures. The Work in Compressed Air Regulations 1958 give rules for compression and decompression. However, more recently improved tables (rules) have been devised which may minimise the likelihood of workers suffering decompression sickness. They are known as the Blackpool Tables and may be used instead of the tables provided in the schedule to the 1958 Regulations. However, permission to do this in Britain must first be obtained from the Health and Safety Executive.

The increased risk of fire calls for the highest standard of tidiness and housekeeping. The adoption of a fire-resistant fluid rather than mineral oil for hydraulic systems should be given careful consideration as the latter is highly flammable.

Burning and welding should only be carried out in the workings where absolutely essential. Even then it should only be allowed when authorised by the person in charge of the site. The work should then be strictly controlled and the following precautions should be taken.

- The bottles of flammable gas should be kept to the very minimum necessary for the task(s) and should be kept in workings for the minimum length of time.
- During the welding or burning operation all flammable material should be removed from the vicinity of the work.
- Any fixed equipment or material below the area of the welding or burning should be protected by a metal tray containing wet sand.
- Throughout the operation a fireman should be in attendance equipped to extinguish any fire and he should remain on duty, observing for the possible outbreak of fire, for at least 30 min after the completion.
- The gas bottles should be carried and stored in suitable protective containers to prevent damage from accidental impact and should

never be placed where rail or road transport might cause damage from impact.

As a rule fire brigades will not provide fire fighting and rescue cover for compressed-air tunnels. It is therefore necessary for the contractor to provide a specially trained fire and rescue squad. The squad must be fit to enter, and be experienced in, compressed air work. Moreover, it is important that the squad remain familiar with the geography of the tunnel.

Except when the man lock is in use the inner door (to the workings) must be left open. Where the pressure exceeds one bar consideration should be given to the provision of decant locks because the man lock may become too hot to be usable throughout the time decompression period.

Failure of the air pressure will cause flooding. This may be due to plant failure or a blow at the face when the air escapes rapidly. For the first possibility it is necessary to have stand-by compressed-air plant; the compressor plant should be under the constant supervision of a competent operator able to switch air supplies in the case of main plant failure. It is a wise precaution if possible to provide air-tight steel, segmental diaphragms at intervals in the crown of the tunnel to form air pockets, where people can survive until the air pressure is restored.

Fire. A good standard of housekeeping is required in free-air tunnels as well as for those driven in compressed air. The amount of combustible materials and flammable liquids should be restricted to that necessary for each working shift. Flammable liquids should be properly stored in sealed screw-capped metal containers away from other combustible materials and away from positions where they may be inadvertently knocked.

Burning and welding should where possible be carried out above ground but when it is necessary in the tunnel only the minimum number of cylinders of the smallest size consistent with requirements should be taken into the workings. In free-air tunnels the cylinders should be carried in and stored as for compressed-air tunnels. For burning and welding operations in free-air tunnels the precautions required are the same as for compressed-air tunnels.

The assistance of the local fire brigade should be sought regarding the selection of fire fighting and protection equipment, including a suitable audible alarm, and arrangements made for the brigade to help in case of fire. A fire and rescue plan should be evolved with the help of the fire brigade.

Safety audits

Regular inspection of the tunnel workings should be carried out by the safety supervisor to ensure that working procedures based on the hazard analysis are being adhered to and updated as necessary. The site manager should take urgent action to remedy any adverse findings.

Sewers

Hazards of sewer work

The hazards for those undertaking the inspection and maintenance of sewerage systems that are still in use, or have been in use, include

- injury from falls
- a sudden increase in the depth and flow of effluent
- dangerous atmospheres
- infection from bacteria.

Falls. Falls may occur at manholes when people are entering or leaving the sewer, or if they slip on the benching at the bottom of the manhole or on the invert of man-entry sewers while walking to or from the workplace.

In addition to the risk of abrasions and fractures from falling, drowning is also a possibility, even in a shallow depth of effluent, if the fall causes unconsciousness. Moreover, there is a particular risk of bacterial infection when the skin is broken by the fall. Injury may also be caused by tools or materials being dropped down a manhole and striking someone at a lower level.

Sudden increase in flow. Conditions in a sewer can change very quickly. A sudden increase in the rate of flow of effluent may sweep workers off their feet and lead to injury or even to drowning. This may occur following heavy rain, which could be either local or some distance from the area where work is being undertaken. It may also be due to high discharges from industrial premises or to the sudden release of effluent which has built up behind a blockage.

Dangerous atmospheres. Dangerous atmospheres may be due to a deficiency of oxygen, or to the presence of toxic gases or flammable or explosive gases. Oxygen deficiency can occur through its absorption by the sewage, or through the replacement of air by carbon dioxide produced by the decomposition of organic matter in the sewer.

Hydrogen sulphide is the most common toxic gas that is likely to be encountered in a sewer, and this is also a product of the decomposition of organic matter. It is not only very toxic but also very explosive. The highly explosive gas, methane, is also likely to be present in a sewer.

Leaks from gas mains and spillages or leaks from petrol storage tanks may find their way into sewerage systems and create flammable and explosive atmospheres. Moreover, discharges from industrial premises often contain substances which release toxic and explosive vapours.

Infection from bacteria. Infection from bacteria may result from skin contact with contaminated effluent. The sources include discharges from hospitals, research laboratories and industrial premises.

A particular occupational risk for sewer workers is skin contact with water that has been contaminated with the urine of rats. This causes Weil's disease which, unless diagnosed and treated early, can be very serious. It is advisable therefore that all sewer workers should advise their doctor of their occupation. Others who enter sewers for any purpose should also advise their doctors that they have done so. Infectious matter may also be introduced into the body through the mouth, nose or eyes, e.g. following a fall into the effluent.

Reducing the hazards

Falls. The main cause of sewer workers falling within manholes is slippery, defective or loose ladder rungs or step irons. Ladders, step irons and landings should be hosed down periodically to clean off all silt, slime or other slippery substances. They should also be checked to ensure that they are secure and not defective; where they are defective the damage should be corrected.

The first person in a working party to descend a manhole should wear a harness with safety line attached which should be kept reasonably taut and controlled by a man at the surface. Until the first person signals from the bottom that entry is safe, no one else must enter except for rescue purposes. Thereafter, the rest of the party should enter with only one person at a time on the ladder. Moreover, in a man-entry sewer, they should move away into the sewer so as not to be directly below the person descending. Similarly, on leaving, only one person at a time should be on the ladder.

Manholes over 6 m in depth should be provided with rest platforms every 6 m and there should always be a rest platform in man-entry sewers just above the soffit level of the sewer. Preferably the platforms should extend over the full cross-section of the manhole and be provided with hinge trap doors. The platforms not only provide a resting place but also limit the distance of a fall and provide protection to those below from falling tools and materials.

All tools, equipment and material must be lowered (or raised) in a suitable container on a rope. They should never be carried by operatives on the entry ladder. When lowering or raising is being undertaken advance warning must be given from the top to allow people in the manhole to stand clear. Moreover, everyone working in sewers or manholes should wear a safety helmet.

When the bottom of the manhole is reached by the first man he should secure a safety chain across the sewer on the downstream side. Furthermore, when the working area is reached, a safety chain should be secured across the sewer, immediately on the downstream side, to prevent workers being swept away should they slip.

When entry is to be made into a sewer, a safety rope, to be used as a handhold, or for the attachment of a lanyard from a safety harness, should be fixed in the sewer between adjacent manholes where steadying support cannot be obtained from the sides or when the effluent is particularly deep.

Sudden increase in flow. Before work starts the person in charge should ascertain whether any industrial establishments have reported their intention to make large discharges into the system. If so, entry should be delayed until the effect on the working area has passed.

Information about weather conditions for the area should be obtained and periodically updated and passed to the men at the surface. If a storm occurs in the vicinity or can be seen some distance away, the top man should signal to those underground to return to the surface immediately. Moreover, where those in the sewer notice an increase in the flow rate, hear the noise of approaching water or a distinct increase in air movement, they should leave the sewer immediately.

Dangerous atmospheres. It is essential to provide good ventilation. This can be done by removing sufficient manhole covers upstream and downstream (at least one either side) of the entry and exit manhole. Before the atmosphere is checked 15 minutes should be allowed. No naked lights, smoking or any means of igniting should be permitted within 5 m of any removed manhole cover or, of course, in a manhole or the sewer.

The atmosphere in both the entry and exit manholes should be tested with appropriate gas detection equipment, and no entry should be made unless safe conditions are indicated. After the working party has entered the sewer the gas detection equipment should be taken with the party and frequent checks made of the atmosphere. Where conditions deteriorate the party must return immediately to the surface. Even when the gas detection equipment indicates safe conditions, if unusual smells are noticed, the party should return to the surface at once.

When entry is made into the sewer, the working party should keep close together and walk slowly and carefully to avoid disturbing the sludge since this may release toxic, flammable or explosive gases. Nylon material should not be worn and any boot studs must be of a non-sparking type.

Infection from bacteria. All cuts, scratches, abrasions and open sores should be protected by waterproof dressings before entry into a manhole or sewer. Moreover, any cuts or abrasions sustained in the manhole or sewer must be cleaned and treated immediately outside the workings. An application of a lanolin-based barrier cream before work starts and following the after-work wash reduces irritation and the risk of infection.

It is important to avoid contact of the skin with sewage and the face must not be touched with the hands. Should a man's face or eyes become irritated he should leave the workings, thoroughly wash and then receive attention. In sewer work, coveralls or sewer suits, sweatbands, gloves, studded thigh or wellington boots must be worn in addition to safety helmets and safety harnesses. Arrangements should be made for drying and cleaning the protective clothing.

At the completion of work it is important that all personnel should thoroughly wash hands and forearms and clean their finger nails in hot soapy water containing an antiseptic. This must be done after protective clothing has been removed and before any food or drink is taken.

Safety at ground level
Safety barriers, cones and road signs must be erected around all open manholes, whether attended or unattended.

Communications
The underground working party must establish and maintain contact with the men at the surface at all times, e.g. by telephone, radio or voice. They should also maintain visual and verbal contact with each other. A check call to ensure that the party below ground is safe and well should be made at intervals not exceeding 3 min from the time when the party leaves the surface. Calls may be verbal, or prearranged signals may be given on whistles or foghorns, e.g. one blast 'are you OK?'; reply, one blast 'yes'. When work is to be carried out along a sewer a man should be stationed below ground at both the entry and exit manholes and the check

call should then include the two men below ground, i.e. surface man (entry) to man below ground (entry) to working party to man below ground (exit) to surface man (exit).

Size of working parties
In no circumstances should anyone enter a sewer or manhole over 5 m deep unaccompanied. Where it is necessary for a party to enter a sewer the minimum size of gang is seven; two surface men and one man below ground at the entry manhole, one surface man and one man below ground at the exit manhole and two in the working party. However, where work in the sewer is within 15 m of the entry manhole support men at the exit manhole are unnecessary. Where the work is entirely within the manhole and its depth does not exceed 5 m the work below ground may be carried out by one person but with two surface men in attendance.

Rescue
The local fire brigade should be informed of the work being undertaken and its location so that they are equipped and ready in case of an emergency. The men at the surface should ensure that they know the location of the nearest working telephone or preferably a portable telephone or car phone should be available.

Rescue apparatus and equipment
The following rescue apparatus should be kept in the charge of the senior man at the surface in close proximity to the entry manhole

- first-aid kit
- two sets of rescue breathing apparatus
- one resuscitator
- one winch (ready for use)
- one escape self-rescue breathing set for each person below ground.

Work in confined spaces

Every year there are a number of fatal and serious accidents because people carry out work in confined spaces on construction sites where the circulation of fresh air is restricted. Confined spaces include trenches, shafts, bored piles, boreholes, tunnels, headings, adits, sewers and drains, underground chambers, pits and ducts, pipelines, tanks, retorts, vessels, vats and silos, furnaces and ovens, box girders and columns, and closed and poorly ventilated rooms, particularly basements.

Hazards of confined spaces
The chief risks in confined spaces are those associated with flammable gases and toxic fumes and vapours and with the likelihood of there being insufficient oxygen to support life. These aspects are dealt with more fully in the sections concerned with excavations, tunnels and sewers (*see* pp. 49, 106 and 112). Reference may also be made to HSE guidance note GS 5 *Entry into confined spaces* (see Appendix 4).

Dangerous concentrations of gases and vapours can arise

- in an excavation, carbon monoxide or carbon dioxide or both from the exhaust of internal combustion engines
- from fissures in the ground, methane gas, especially in carboniferous areas, refuse dumps and reclaimed land with domestic and industrial waste; it can be flammable and highly explosive when mixed with air
- nitrous fumes from the use of explosives
- in a vessel or container from dangerous chemicals remaining from a previous use
- when sludge containing dangerous chemicals is disturbed by cleaning operations
- in live sewers from industrial waste containing dangerous chemicals or biological action in rotting vegetation, for example
- when welding or flame cutting
- when using solvents or applying adhesives, e.g. waterproofing operations
- due to oxygen enrichment of the atmosphere by operations using oxygen, e.g. oxy-propane cutting.

Oxygen deficiency can occur in pits and headings, for example, as a result of the displacement of oxygen by carbon dioxide which has seeped in from the surrounding ground, particularly in chalk and limestone. Oxygen deficiency can also occur in steel box girders and in other confined spaces in steel structures where oxidation of the steelwork has occurred.

Safe systems of work in confined spaces

It is recommended that a permit-to-work system is followed for all operations that require entry into confined spaces (*see* HSE guidance note GS 5 *Entry into confined spaces* and chapter (6). No staff or operatives should be allowed to enter a confined space if they are physically or mentally unsuitable. Rescue from confined spaces is difficult enough without the added burden of chronic disability.

Construction work in confined spaces is covered in Regulation 21 of the Construction (General Provisions) Regulations 1961. It requires the provision of adequate ventilation to maintain an atmosphere fit to breathe and to render harmless all dust, fumes or other impurities injurious to health. If there is a likelihood that the atmosphere is not safe, no one must enter the confined space until the atmosphere has been tested by a trained competent person and until that person is satisfied that the air is safe to breathe. Atmospheric testing may be carried out by portable oxygen meters, gas detector tubes and explosimeters. Guidance is given in HSE guidance note EH42. A safe atmosphere may be provided by the introduction of fresh air naturally or by forced ventilation and extraction. This may be by compressed air, either from cylinders or a compressor, by a blower fan and trunking and/or by an exhaust fan and trunking.

Where rigid trunking is used the fans can be reversed to dispel noxious fumes, for example where explosives are used in tunnelling. In all cases it is important that fans and trunking are maintained in good condition. Where the provision of fresh air is impracticable, people entering should wear breathing apparatus. The apparatus may be portable with the compressed air cylinder carried on the back harness, or otherwise the cylinder is located outside the confined space and the breathing apparatus is connected to it by airline.

Working precautions

The precautions to be taken when working in excavations, tunnels and sewers are covered on pp. 49, 106 and 113. It is essential that trained personnel should be in attendance whenever someone is at work in a confined space and it is important to ensure that accident assistance is immediately available. Those inside the confined space must keep in verbal communication with those outside or else rope signalling may be used. A portable telephone or two-way radio is desirable as an alternative. Rescue equipment must be provided and should include

- two safety harnesses, with an adequate length of safety line
- safe hand torches or cap lamps; an audible alarm
- a rescue team with breathing apparatus, first aid resuscitation stretcher and fire-fighting equipment.

Working on contaminated sites

Increasingly, civil engineers, builders and demolition workers are required to design or construct works on sites that are contaminated by waste products. Contamination occurs from domestic and industrial refuse and from processes involving the production and storage of chemicals.

Many chemicals are injurious to health if absorbed into the body, some show immediate effects and others are serious only if built up over a period. Absorption may be through the skin, inhalation or ingestion. Examples of chemical hazards and precautions to be taken are given on pp. 132 *et seq*. Sites where contamination might be expected include

- asbestos producers
- chemical plants and depots
- dockland areas
- explosives factories and depots
- gas works
- landfill sites of domestic and industrial waste
- metal smelting and refining plants
- metal treatment and finishing works
- mines and quarries
- oil production and storage depots
- paints and graphites factories and depots
- railway yards
- scrap yards
- sewage works
- steelworks
- tanning and associated trades.

The distribution of contaminants is often random and not always identified by trial pits and boreholes. The contaminants may migrate through the ground with ground water and their chemical compositions may change by contact with one another. Some organic substances decompose, producing methane and sometimes hydrogen sulphide (see pp. 49, 50, 102, 106, 115 and 134).

Identification and remedial action

The likelihood of contamination on a site may be confirmed by reference to clients' records, local authority records, old maps, etc. Soil samples should then be obtained, taking care not to expose to risk those carrying out the task, and sent to a suitable laboratory for analysis and estimation of the concentration of harmful chemicals present. No construction or demolition work that might endanger people on or near the site should be allowed until the contaminants have been identified and precautions taken.

For the development of contaminated sites it may be necessary to remove some or all of the contaminated material. Disposal of contaminated material comes under the Control of Pollution (Special Wastes) Regulation 1980, which is administered by the waste disposal department of the local authority. This will influence the cost of disposal and must be taken into account in the preparation of tenders. There must be no encouragement of fly-tipping. Where removal is considered unnecessary or undesirable the site may be covered by a blanket layer of granular material, the thickness depending on the protection required, the topography and future construction on the site, e.g. piles, foundations and drain or service trenches.

General advice on the problems of contaminated sites may be obtained from the Interdepartmental Committee of the Redevelopment of Contaminated Land, Department of the Environment. Advice may also be obtained from the local office of the Health and Safety Executive. A number of consultants specialise in the monitoring, measurement and control of contaminants and are able to advise on removal and replacement or on other site surface preparation.

Site precautions

People at risk include surveyors, site investigation teams, construction workers, demolition workers, visitors to the site and others in the vicinity. Suitable warning notices must be displayed and secure fences erected.

Workers on the site should be provided with

- a detailed work plan
- information on the hazards and precautions to be taken, and advice on personal hygiene
- clean overalls, gloves and other protective wear as necessary
- a site hut with toilets, washing and changing facilities and a suitable eating place.

Work over water

For work over or in the vicinity of water, the ever-present risk of falls is compounded by the risk of drowning. In the prevention of accidents the approach recommended here is to reduce the possibility of falls and to provide rescue facilities and equipment.

Prevention of falls

Working platforms, places and access gangways to them should be properly constructed to the widths required by the Construction (Working Places) Regulations 1966. Rigid guard rails and toe boards should be fixed at the correct height at all edges from which falls are possible. This also applies to the edges of openings within the structure through which falls are possible, e.g. the uncompleted formwork of a bridge deck.

The decking to workplaces and gangways should be close boarded and effectively secured to prevent displacement by strong or gusty winds or by rising water from tide or wave action when the decking is close to the water surface. Ladders for access should be sound, securely fixed and of sufficient length to enable people to get on and off without difficulty.

Tripping and slipping hazards must not persist unchecked. Unnecessary materials, tools and equipment should not be stored or left at working places or on gangways. Materials required for the work should be restricted to those for the immediate tasks in hand and should be neatly and securely stacked. The decking of working places and gangways in marine work can become wet, slimy and greasy. The surfaces should be regularly cleaned, e.g. to prevent slipping. Sand that has been put down for cleaning oil spillages should be swept up. During periods of frost or snow, cleaning and salting should be carried out and the salt should be swept up before other work is carried out.

Non-slip footwear should be worn by all men using platforms and gangways. Where the provision of safety nets immediately below the working areas is impracticable, the use of safety harnesses with lanyards continuously attached to a secure anchorage is advisable.

Whenever visibility is not good, a high standard of artificial lighting must be provided for work areas and accesses. This should also illuminate the surface of the water adjacent to site operations.

Rescue equipment

All people working over water should be provided with and encouraged to wear a brightly coloured life jacket, or other buoyancy aid capable of bringing a man to the surface, even if unconscious, in a face-up position. It should have self-igniting lights to facilitate easy identification in poor visibility. Buoyancy aids should comply with BS 3595: 1969[14] or be of a type approved by the Ship and Boat Builders National Federation.

Board of Trade approved life-buoys with lifelines should be provided at intervals along the site and positioned conveniently for use in an emergency. In addition, lightweight emergency lifelines should also be readily available. They can be thrown up to 40 m with reasonable accuracy and are easily recoverable for rethrowing when necessary.

Grab lines should be attached at intervals to structures or floating plant. A marker float should be attached at the free end. It may be prudent to provide additional grab lines downstream of the operations. Checks

Fig. 4.32. Uncontrolled collapse of multiple-span brick arch bridge owing to incorrect demolition (courtesy HSE)

should be made at the beginning of each shift to ensure that rescue equipment is present and in good order.

Rescue facilities

It is essential to provide arrangements for the prompt rescue of anyone falling into the water. Moreover, at every place where work is being carried out on the structure at least two men should be employed so that, should one fall, the other can raise the alarm.

A suitable rescue boat manned by two experienced boatmen wearing life jackets and who are also qualified first-aiders must continuously patrol the site of operations while work is in progress. The boat should be fitted with grab lines, at least one life-buoy or lightweight emergency lifeline, a boat hook, anchor and line and a baler. Two-way communication with the site office and the working level should be provided. Where work is to be carried out in darkness or at dusk, a swivelling searchlight mounted on the boat is essential. For rescue purposes it is useful to keep on board a pole with a hooked end.

Water transport

Where people are conveyed by water to and from their place of work the boats provided must meet the requirements of the Merchant Shipping Act. Where the craft is intended to carry more than 12 people in addition to the crew, it must be surveyed at least once a year by a Board of Trade surveyor and certified as being seaworthy, safe and possessing the required equipment. This includes inflatable life rafts, life jackets, life-buoys, distress signals and fire extenguishers. There are no rules for craft smaller than for

12 passengers but, of course, as with larger boats, the crew must be competent and experienced boat handlers.

Demolition

Demolition is one of the high-risk activities of the construction industry with a fatal and major injury accident incidence rate about 17 times that for the whole of the construction industry. Approximately 10% of all fatal accidents each year in the construction field occur in the demolition sector. Detailed guidance on acceptable standards and procedures in demolition is contained in BS 6187: *Code of practice for demolition*[15] and in HSE guidance note GS 29 *Health and safety in demolition work* (*see* Appendix 4).

Demolition is a very skilled process and, to be carried out safely and correctly, requires a great deal of knowledge of building construction and expertise in demolition methods. It is dangerously wrong to think otherwise (Fig. 4.32).

Key to safe demolition
For any demolition operation (including partial demolition such as in refurbishment projects) to be undertaken safely and efficiently, not only is a careful preliminary investigation of the structure necessary but also the results of each step in the demolition process must be accurately forecast. The steps to be followed are listed below.

Pre-survey inspection. A pre-survey inspection of the structure should be undertaken to ascertain

- the type of construction: this should include the contribution made by key elements to stability, e.g. by roof members, floors, walls and structural frameworks, etc.; detailed structural features should also be identified such as the counterweighting or holding-down method for cantilevered elements, tie bars — particularly where used in arch floor construction — the direction of main bars in reinforced concrete floors and the type of stressing in prestressed concrete members
- the state of the structure: the condition of structural members and of the fabric; damaged areas, previous repairs and alterations should also be noted
- the need for temporary shoring or other support: this is particularly relevant for attached buildings, arch construction and refurbishment
- special features: this would include the presence of dangerous materials or chemicals, e.g. asbestos, lead paint or dust, flammable residues and vapours in storage tanks or process vessels and contaminated ground
- the position of services: it is important to identify the whereabouts of gas, electrical and water services and to ensure that they are disconnected or isolated before demolition work starts.

Demolition method. The pre-survey should have identified the problem areas and provide all the information necessary to determine the method of demolition to be used. The object should be to limit the number of people at risk and thus to adopt where possible methods which make it unnecessary for men to work at heights.

Moreover, as already indicated, it is very important to forecast the result of each step in the demolition process and thus to determine the precautions necessary to prevent premature collapse causing injury to personnel.

Method statement. When the method of work has been determined it should be set down in detail in a clear and understandable form. It should include

- the method to be adopted and the sequence of operations, the plant and equipment to be used and the details of working platforms and access routes required
- where any pre-weakening is to be carried out, the precise details of how this is to be done
- arrangements for the protection of the public and also for all personnel employed on the site; the arrangements should include the provision of protective fans or nets, and hoardings or other provision for preventing public access to the building being demolished; for site personnel the arrangements should cover the operation and maintenance of all plant and the provision and use of the necessary personal protective equipment
- arrangements for the disconnection or isolation of services
- in detail, the methods to be used for dealing with dangerous materials and hazards chemicals, e.g. asbestos and residues in storage tanks
- the identity and responsibilities of the person in charge of operations and the responsibilities and duties of people in his charge who will control certain aspects of the demolition work.

Site supervision. Obviously not every man engaged in demolition work will understand the structural significance of the various component parts he has to demolish and therefore demolition workers will need close supervision and guidance from someone skilled in the work if they are to proceed in safety.

It is important that supervisors have a considerable knowledge of building construction as well as demolition methods. They must always be experienced and competent in the particular operation being undertaken and be able to provide step-by-step guidance for each stage of the operation.

Safe clearance zones. A restricted zone for all people should be maintained around the structure to be demolished. This should be a clear distance of 6 m or half the height of the structure (whichever is the greater) from the point of fall so that the debris can fall freely. If this is not possible then the debris must be lowered in skips or by chute.

When mechanical plant (e.g. cranes, grabs, pusher arms or lifting machines for balling purposes) is employed on site, a minimum restricted zone 6 m wide must be maintained from the face of the building. When the

Fig. 4.33. Unacceptable method of demolition (courtesy HSE)

demolition method is pulling by wire rope, the restricted zone should be three-quarters of the distance between the winch and the structure to be demolished on either side of the rope and forward of the winch (or tracked vehicle) used for pulling. The appropriate restricted zone when tall structures such as chimneys are being felled should be an area of radius $1\frac{1}{2}$ times the height of the structure measured from the base.

Methods of demolition

The danger in demolition work varies with the method employed, which itself has a bearing on the number of men at risk and their working time at heights and in and around the structure being demolished. Methods of demolition may conveniently be divided into three categories, hand, mechanical and explosive.

Demolition by hand. Demolition by hand is progressive demolition, carried out piece small with hand tools such as sledge hammers, jack hammers, concrete breakers and cranes for lifting and lowering purposes. It is carried out in reverse order to the construction process, i.e. the roof covering is taken off first, followed by the roof trusses and upper ceiling and the building is then demolished storey by storey.

When this method is used properly protected working stages should be provided where necessary. Standing on the wall being demolished is unnecessary and should not be permitted (Fig. 4.33).

When operatives are expected to work at unprotected edges such as when breaking up a floor slab harnesses should be worn with appropriate safety lines suitably anchored. Where parts of floors have been removed

124   **CONSTRUCTION SAFETY HANDBOOK**

Fig. 4.34. Use of purpose-made man-riding skip to provide a safe working place for demolition of a high structure (courtesy HSE)

to allow debris to fall through, access to the areas concerned should be prevented or guard rails and toe boards erected at a distance from the edges to protect workers from falling debris. Similarly, any holes in floors in the building under demolition should be protected by guard rails and toe boards to prevent people falling.

Structural collapse has occurred where floors have been overloaded by the accumulation of debris from parts of the building already demolished. Debris should not be allowed to build up, but should be discharged regularly by chute or skip or by dropping it to ground level. Where debris is dropped internally the build-up at ground level must be limited to avoid imposing excessive loads on the walls. No debris should be allowed to fall while clearing operations are in progress.

Where scaffolding is used to provide a working platform care is needed to ensure that sufficient ties are provided at the lower levels before demolition (and hence removal of ties) at the higher levels. Work on roofs should only be carried out from properly constructed platforms or crawling boards. Safe access for demolition workers is sometimes difficult to provide. Man-riding skips (Fig. 4.34) may be used.

Mechanical methods of demolition. Mechanical methods of demolition include the use of a demolition ball or pusher arm; inducing deliberate collapse; and pulling by wire rope; some examples are given below. The great advantage of mechanical methods is that they considerably reduce the number of people at risk, i.e. the number of people required at height on the structure. Sometimes it is necessary to demolish by hand until the structure has been reduced to a height convenient for one of the mechanical techniques to be used.

For demolition by *demolition ball* a swinging weight is used suspended from a lifting appliance, to demolish the structure progressively. The method should only be carried out by skilled operatives. Before a machine is used for balling operations the advice of the machine manufacturer should always be sought to ensure that the machine is suitable for the work. Great care is needed to avoid overstressing the lifting machine and it is preferable for the ball to be dropped vertically or swung in line with the jib. Where a slewing action of the jib is used the angle of slew should not exceed 30° and the acceleration of slew and rate of checking should not be excessive — advice should be sought from the machine manufacturers.

The angle of the jib should not exceed 60° to the horizontal and its head should never be less than 3 m above the part of the building being demolished. When this technique is used the floor areas at all levels should first be reduced by 50–75% to allow for falling debris.

For demolition by *pusher arm* a steel attachment is used, fixed to an extended jib arm on mechanically operated mobile plant. It exerts a horizontal thrust similar in technique to a battering ram and is used to demolish walls.

The equipment should be on firm and level ground and used from outside the building. The cab needs to be protected against the impact from falling debris and the windows should be made of shatterproof glass. The application of the pusher arm to the wall being demolished should never be more than 600 mm below the top of the wall.

In the demolition method of *deliberately induced collapse* key structural members are removed, causing the whole or part of the building to collapse. The method requires a high degree of planning and the use of expert engineering advice. Where partial demolition of a structure is to be carried out by this method consideration must be given to the effect it may have on the section to be left standing. Where there is the slightest doubt about the stability of the remaining part the method should not be used.

Collapse is usually brought about by pre-weakening the key members and their final removal by means either of explosives or the method of pulling by wire rope and winch (or suitable vehicle). The most dangerous part of the method is the pre-weakening when, by the very nature of the exercise, operatives are working on and below a structure close to instability. The planning and weakening operation must be such that wind loads, likely vibration from work being carried out in the vicinity and any impact loads (e.g. those which might be caused by the operatives carrying out the work either deliberately or accidentally) must not tip the balance to instability.

Structural steel elements are usually weakened by cutting with an oxy-acetylene torch and the type (pattern) of cut is important. Fatal accidents have occurred through incorrect cutting and overweakening of the supporting pattern. With masonry or reinforced concrete buildings a series of rectangular holes is cut towards the base so that the remaining building is effectively held up on a number of masonry (or concrete) columns — between the holes. The holes are cut with hand tools — jackhammers etc. — and care must be exercised to ensure that falling masonry from the top of the holes does not cause injury; temporary support may be necessary, especially on fairly thick walls where the operatives may have to stand within the cut area.

Before collapse is implemented all personnel should retire to a safe distance, and where explosives are used be protected from possible flying debris by a suitable barrier.

In addition to its use in the final stages of the deliberate collapse method of demolition, *pulling by wire rope* may also be used for parts of buildings such as chimney stacks and walls without pre-weakening, other than perhaps partial removal by hand of the adjacent walling and ceiling or floor.

The wire rope is fixed around the wall (or part of the building to be demolished) at one end and the other end is fixed to the pulling winch or tracked pulling vehicle. The main dangers are that people, in particular the winch operator or driver, may be struck by falling material; the rope may sever and snake, hitting either the operator or people in the vicinity; where the method is unsuccessful and the part being pulled remains standing, it and adjacent parts will have been weakened and may be in a dangerously unstable state; moreover, there is the danger of the pulling vehicle overturning if it lacks weight and power for the job.

The method of pulling by wire rope should not be used on buildings (or parts) that are more than 21 m in height. The length of the rope should be such that the winch operator or driver is positioned at a horizontal distance from the building of not less than twice the height of the highest part being pulled. Measures must be taken to ensure that people are prevented from entering the restricted zones indicated in the method statement, and the winch operator or driver must be protected from falling or flying debris by a suitable substantial cabin.

Steel wire ropes should be used, in good condition (undamaged) and of minimum circumference not less than 38 mm. However, the correct size of rope will depend on the load imposed by the operation. The rope should be examined by a person competent and experienced to judge the quality, condition and adequacy of ropes for the work, before use, at least twice daily and again before further use where damage during use is suspected. The rope should be protected where it is wound around sharp edges to prevent high stress points developing and to reduce wear.

The winch or pulling vehicle should be aligned in the direction of pull. Where a winch is used it should be well anchored and where a vehicle is used it should not lift off its tracks — if this starts to occur the operation must be abandoned.

Fig. 4.35. Safe mechanical demolition showing use of a machine-mounted impact hammer (courtesy HSE)

When the operation starts the load should be gradually applied — snatch loading should never be used. Following the application of full pulling load should the structure not collapse an alternative mechanical method must be used. In no circumstances should hand demolition be reverted to nor should any operatives be allowed on to or even to approach the structure. The most suitable alternative methods are the use of a demolition ball or pusher arm.

Other mechanical methods include a number of specially designed hydraulically operated devices for demolition without the need for operatives to work at heights. They include machine-mounted impact hammers (Fig. 4.35), power shears (an alternative to oxy-acetylene cutting)(Fig. 4.36), grabs, nibblers and concrete bursters (also gas-expansion concrete and masonry bursters). The main dangers are from falling material, which is overcome by the maintenance of restricted zones for personnel and cab protection for the operator where applicable; and overloading of the machine causing collapse of the arm or even overturning of the machine, which is overcome by the use of trained and experienced operators.

Use of explosives in demolition. Given the necessary expertise demolition by the use of explosives is possibly the safest method of demolition because the fewest people are at risk. Only specialists with the required experience and training should plan and control this type of work.

The main danger is from flying debris; accidents may be averted by

- use of the correct type and charge weight of explosive: this is not usually a problem when an explosives expert is in control

Fig. 4.36. Safe mechanical demolition showing power shears for steelwork demolition (courtesy HSE)

- use of adequate blast mats to reduce the effect of air blast and flying debris
- the provision and maintenance of a restricted zone around the operational area, which should be decided by the experienced person in charge of the operation; before the charge is fired everyone except the shotfirer should be excluded from the area and not allowed back until the all-clear signal is given; the shotfirer should be protected from flying debris by a suitable barrier
- the establishment of a firing timetable together with an audible and visual signalling system to alert everyone, both site personnel and the public, of the intention to start firing; the visual arrangements should include the use both of red flags (to indicate danger) and notices at intervals along the border of the restricted zone and about the site indicating the work to be undertaken, the dates and times of firing, the audible warnings to be given before firing and the significance of red flags flying
- use of sentries: before the charge is fired sentries should be posted at intervals around the perimeter of the restrictive zone and at all road and track access points to prevent entry.

Other precautions which need to be taken to prevent accidents when explosives are used in demolition operations are as follows.

- Correct storage and transportation of explosives and detonators is important: in particular, detonators should be kept and carried separately from explosives until they are required for use and should only be handled by those authorised to do so.
- Neither matches nor lighters should be carried by people handling explosives or detonators or by people in or close to storage places except when needed for firing a charge.
- Explosives should not be handled or kept near the path of vehicles or other mobile plant, where there is danger from falling material; nor should explosives be handled or kept close to sources of heat. Electric detonators should be kept well away from electric welding equipment, electrically operated machinery and vehicles with starter motors, generators and batteries, etc., and from radio and radar equipment.
- Except for the delivery of explosives all vehicles should be prohibited from entering the blast area.
- Electric detonators should only be carried around the site in boxes made of non-conducting material.
- The number of detonators used against those issued should be recorded and the detonator box kept locked until the detonators are needed.
- After the charge has been fired the all-clear should not be given until the shotfirer and person in charge of the explosives operation have conducted an examination of the area and satisfied themselves that it is safe, particularly after a misfire. Where a charge fails to explode no approach should be made for at least five minutes.
- All mobile plant should be moved to a place of shelter (from flying debris) or to a point outside the restricted zone before firing occurs.
- After the charge has been fired, should the structure remain standing all personnel must be prevented from entering the restricted zone until an assessment has been made; the competent person in charge of the operation, with the assistance of other experts (e.g. a structural engineer), as necessary, should determine the further measures required for complete demolition. Moreover, they must assess what (if any) safety precautions need to be taken, e.g. the erection of barriers with notices to prevent access.

Further guidance on explosives is given in BS 5607: *Code of practice for the safe use of explosives in the construction industry.*[16]

Health hazards from demolition
Health hazards from demolition processes arise from

- inhalation of toxic gasses, vapours, fumes and dust
- ingestion of corrosive or toxic liquids and dust
- reaction with or absorption by the skin of corrosive, toxic or irritant dusts, powders, liquids and chemical substances

Fig. 4.37. Use of good protective clothing and breathing apparatus during asbestos removal (courtesy George Wimpey and Co. Ltd)

- exposure to high noise levels
- exposure to vibration.

The hazards should be identified in the method statement and the method of work detailed on how the dangers will be avoided and the risks overcome.

Where there is danger of inhalation of toxic gases etc. a method of work should be devised to minimise the generation of dust and fumes, e.g. in the case of dusts by damping with waterspray and by containment with chutes, covers, etc. Where the risk from dust or fumes cannot be completely eliminated, appropriate personal protective breating equipment must be used by all those exposed to the hazard (Fig 4.37).

It is also important to ensure that airborne contaminants are prevented from escaping beyond the site and affecting the general public. Environmental monitoring should be carried out in the working area and at intervals around the site boundary where toxic dust and fumes, and excessive nuisance dust or both are likely to be generated.

With regard to the danger from ingesting and absorbing toxic substances, not only should personal protective equipment be used but also it is most important that a high standard of personal hygiene is practised. Thus adequate washing facilities must be provided at the site.

With regard to the danger from exposure to high noise levels, operatives should not be subjected to continuous noise levels over 90 dB(A). Where noise levels are higher than 90 dB(A) or where peak levels exceed 200 Pa ear protectors should be supplied and worn (*see* p. 136). The most likely sources of high noise levels are concrete breakers and compressors.

Initially, noise reduction by engineering means (e.g. surrounding concrete breakers with sound-absorbing bags) should be carried out to reduce the noise level to as low a level as practicable before the use of ear protectors is resorted to. Moreover, exposure time should be kept to a minimum by means of job rotation for operatives.

With regard to the danger from exposure to vibration, concrete breakers and pneumatic drills are examples of types of plant, widely used on demolition work, which may give rise to vibration white finger. The use of gloves and the device of restricting exposure by means of job rotation may reduce the risk of serious disability.

Examples and sources of toxic dust, fumes, etc. are as follows.

- *Asbestos*-containing material has been used for both thermal and acoustic insulation purposes. It may be found around hot vessels and pipelines, sprayed on the underside of roofs, and on walls and columns of buildings and structures. It may also be found in sheet form, i.e. asbestos cement, as roof covering and partition walls and in flooring material. All demolition operations will create toxic asbestos dust and it is important to realise that no level of exposure, for any type of asbestos, has been accepted as being safe.
- *Lead* dust may arise from the cold cutting of elements covered with lead-based paint or when metallic lead is handled. Lead fumes may arise from the flame cutting of metalwork painted with lead-based paints for example, and from the dismantling of tanks containing lead-based petrol.
- Flame-cutting plant may contain chemical deposits of *zinc* or *cadmium* and steelwork may have been painted with *zinc* or cadmium paint.
- *Silica* may occur in natural stone (e.g. sandstone), some bricks and concrete aggregate, and any method of demolition of structures constructed from these materials, will give rise to dust containing silica.
- *Ammonia, chlorine* and *aniline* are examples of gases and fumes that may be released during the demolition of refrigeration and chemical plant unless they have been properly purged before the dismantling demolition.
- Various *toxic sludges* may be encountered during demolition of virtually any storage tank forming part of an industrial process.
- All storage tanks and reservoirs in industrial premises may contain harmful products such as *corrosive liquids*.

A number of HSE guidance notes on health hazards that may be associated with demolition are listed in Appendix 4.

PART 2. HEALTH HAZARDS

From chapter 1 it has been seen that occupational health hazards in construction work may be grouped under chemical, physical and biological headings. For the particular health hazards associated with demolition work see also p. 129 *et seq.*

Chemical hazards

Dangers from chemicals may arise from contact with the skin, inhalation or ingestion (swallowing).

Contact with the skin

Dermatitis is perhaps the most common occupational skin disease, and can be caused by contact with a number of materials commonly used in the construction field. Cement, particularly when wetted, is an example and if it contains chromates the danger is increased. The slaked lime content in cement is considered to be responsible for cement burns on the legs and feet of operatives where concrete has spilled into their wellington boots.

Certain epoxy resins used in grouts, seals and adhesives are another cause of dermatitis, as are acids, alkalis, solvents, thinners, paints, varnishes, acrylic and formaldehyde resins, brick and stone dust, pitch, tar and bitumen.

To avert this hazard, it is clear that skin contact with chemicals should be avoided. Careful handling, including the prevention of splashing, is called for, together with the use of closed stout containers for distributing, and appropriate tools for mixing and applying the products. Protective clothing should be worn, in particular suitable gloves and safety eyewear. Barrier creams on the hands and forearms may be of use but care is needed to select the correct product for the particular operation.

Personal cleanliness is most important and hands must be clean before eating, smoking or using the lavatory. Gloves should be thoroughly washed before they are removed. An individual may become sensitised (i.e. allergic) to a particular substance by repeated exposure and even after medical treatment the dermatitis will recur on further exposure to the substance.

Inhalation of harmful chemicals

The inhalation of certain dusts, fumes, gases and vapours may give rise to a variety of risks to health, including respiratory disorders, poisoning, asphyxiation and cancer. HSE guidance note EH 40 (*see* Appendix 4) gives occupational exposure limits for many airborne contaminants. It is revised annually. Some of the more common substances are now discussed.

Asbestos dust. All types of asbestos dust are dangerous and breathing them may lead to asbestosis, which is a scarring of the lungs causing shortness of breath. Other hazards are lung cancer and mesothelioma, a rare cancer of the lung envelope. Asbestos is present in a number of products used in construction processes, e.g. asbestos cement roofing sheets, decking, cladding sheets, rainwater and soil pipes. Their cutting, handling and breaking gives rise to this dust. It is also found in asbestos insulating board and in sprayed coating for insulation and fire protection purposes. (This process using asbestos-containing material is prohibited under the Asbestos (Prohibitions) Regulations 1985.) However, stripping out the coatings continues to be an operation very hazardous to health. The Asbestos (Licensing) Regulations 1983 are intended to prevent employers and the self-employed carrying out work that involves disturbing asbestos insulation and coatings unless they are licensed to do so by the Health and Safety Executive.

Wherever possible safe substitute materials should be used for the building process. Otherwise, exhaust ventilation equipment, approved respirators and appropriate protective equipment must be used when handling and manipulating asbestos or asbestos-containing materials. For operations such as drilling, cutting and cleaning up, the area should be dampened and the operation carried out carefully to minimise the creation of dust.

A particular risk associated with dust on protective clothing is that the dust may be breathed in by anyone handling the garments, e.g. for cleaning. Immediately after use, protective clothing should be carefully placed (so as not to cause dust) in a sealed container appropriately marked 'asbestos-containing clothing' and sent for cleaning by cleaners competent to carry out the work. Those intending to work with asbestos should refer to the Control of Asbestos at Work Regulations 1987 and guidance notes published by the Health and Safety Executive (*see* chapter 8 and Appendix 4).

Silica dust. Silica dust is another commonly found dust which causes damage to the lungs. The lung scarring is called silicosis and causes breathing difficulties similar to asbestosis. Free silica is present in many rocks, aggregates and sands and danger arises from carrying out processes involving such materials. The processes may include tunnelling through rock, slate or ground containing flint nodules; the abrasive cleaning of stone structures, grinding, cutting and working granite and other stone; rock drilling and blasting. The precautions to be taken include standard dust-control methods such as exhaust ventilation, wetting down and, where appropriate, the use of protective breathing equipment and dust-proof protective clothing.

Carbon monoxide. Carbon monoxide is one of the most dangerous gases likely to be encountered on a construction site. It is odourless and, therefore, easily breathed by the unsuspecting. Headache, nausea, dizziness and loss of co-ordination may result from low levels of exposure, whereas high levels lead to rapid death. Moreover, regular low exposure may lead to heart disease.

Sources include internal combustion engine exhausts, poorly maintained space heaters (with incomplete combustion of the fuel), coke braziers and carbon dioxide welding. The risk is particularly high in confined working areas such as tunnels and excavations. In all cases the need for adequate ventilation is paramount to avoid danger.

Where carbon monoxide is likely to be present then appropriate monitoring of the atmosphere must be carried out.

Carbon dioxide. Carbon dioxide occurs naturally in chalk and limestone areas. High concentrations in confined spaces result in oxygen deficiency and very dangerous atmospheres. Numerous fatal accidents have occurred when unsuspecting workers have entered manholes in surface-water drainage systems or boreholes in such areas.

In all shafts, wells, adits, tunnels and excavations in limestone and chalk areas it is particularly important before each entry and continuously during all working time that the atmosphere should be tested for high concentrations of carbon dioxide and for oxygen deficiency. A further most important check is the barometer reading. If it falls during the time workers are in the excavation the atmosphere may change lethally, and therefore immediate oxygen-deficiency testing should be undertaken. Again, the necessary precautions are adequate forced-draught ventilation and atmospheric testing before entry and throughout the working cycle. Where there is a need to enter a dangerous or unknown atmosphere, appropriate breathing apparatus should be worn.

Hydrogen sulphide. Hydrogen sulphide in high concentrations can cause death, and at low levels, irritation to the eyes, nose and throat, dizziness and headache. It may generally be found in sewage, other decomposing organic materials and excavations in made-up ground. The necessary precautions are similar to those indicated for carbon dioxide. Workers carrying out maintenance in existing sewers should be particularly on the alert and should not enter manholes etc. without carrying out checks of the atmosphere, and should then undertake periodic checks while working underground.

Nitrous fumes. Nitrous fumes are very toxic: exposure to high concentrations may prove fatal and exposure to lower concentrations may result in bronchitis. A common source results from the use of explosives in tunnelling and excavating, for example. The confined space of the tunnel and excavation is perhaps the area of greatest risk. The main precaution is, if possible, to ensure that personnel are removed from the danger area during blasting and firing operations and that they stay away until the fumes have dissipated.

Cadmium. Breathing cadmium fumes would almost certainly result in cadmium poisoning, which occurs suddenly, causing drowsiness, vomiting, loss of muscular control and possibly death. Cadmium fumes are released during cutting, welding and brazing operations with cadmium-plated steel.

The main danger occurs when the operations are carried out in confined spaces such as tanks and fabricated structures, and the necessary precaution to take is the provision of good ventilation. It will often be necessary to provide local exhaust ventilation, especially in confined spaces, or to ensure the use of air-line breathing apparatus.

Lead. Lead poisoning may arise from breathing either lead dust or lead fumes and results in fatigue, anaemia, colic or wrist-drop. The cutting and burning of old structures covered in lead-based paint releases lead dust and fumes and therefore the highest risk area is probably demolition work.

Where possible the process should be segregated and appropriate (possibly local exhaust) ventilation provided. Good protective clothing should be worn (i.e. a type that does not hold dust), and in some cases respiratory protective equipment should be used. An essential precaution is personal hygiene — hands should be thoroughly washed before workers eat, drink or smoke.

Overalls or other protective clothing should be removed before meals are taken and not stored or taken into areas where food and drink are consumed. It is also important that users should wash immediately following the removal of all protective clothing. Similar standards of personal hygiene should be followed by operatives before they smoke etc.

Welding fumes. In welding operations the parent metal and its coatings, together with the weld metal and rod coating, release a variety of complex fumes which, if breathed, may give rise to metal-fume fever — a flu-like illness. The necessary preventive measure is the provision of local exhaust ventilation and, for short periods, respiratory protective equipment of the type that supplies fresh air. Moreover, as with lead, good personal hygiene is very important.

Zinc. Zinc fumes are evolved from the welding, brazing and flame cutting of galvanised steel, and breathing them may cause zinc-fume fever. Exhaust ventilation and respiratory protection are the precautions to be taken if this hazard is to be avoided.

Solvents. Various types of solvent are contained in a wide range of products used in the construction industry, e.g. adhesives, sealers, paints, lubricants and lacquers. Not only may they cause dermatitis through skin contact but they may also be a source of vapours which, when inhaled, can cause a variety of illnesses including headaches, dizziness and vomiting, and, in severe cases, unconsciousness preceding death. They are particularly hazardous when used in confined spaces (e.g. basements) because of higher concentrations and poor natural ventilation. A good standard of ventilation and appropriate protective clothing should be adopted to minimise these hazards.

Ingestion of harmful chemicals

The swallowing of harmful chemicals by mistake or through lack of personal hygiene is an ever-present risk in construction and demolition activities. The hazards and precautions against the ingestion of harmful chemicals are largely covered by the commentary above on dangers arising from contact with the skin and inhalation. The promotion of good personal hygiene, the provision of correct protective clothing, equipment, toilet and washing facilities, accompanied by training and supervision, are also essential to prevent the transmission of chemicals to the mouth from hands, arms, gloves, etc. In addition, there is a need for clear correct labelling of harmful substances with special attention paid to liquids not in their original containers, to warn against accidental swallowing.

Physical hazards

In the second main category of hazard danger results from the general environment experienced by the workers or from their particular occupation.

Cold

Construction workers may be exposed to conditions of extreme cold and to biting winds accompanied by rain over long periods. Such prolonged exposure without adequate protection may have serious effects on their health. The effects may range from a dulling of mental faculties and slower muscle reaction to possible bronchitis and arthritis. Furthermore, errors and accidents are likely to increase when the hands of operatives are particularly cold.

The first line of defence against the cold is to wear suitable protective clothing, supplemented by sensible undergarments and warm footwear. Moreover, if possible, overall general protection to the work area should be provided, e.g. the use of reinforced plastic sheeting in the form of screens. The provision of warm accommodation where hot drinks and food are available is essential.

Heat

Although not as prevalent in Britain as in countries with much hotter climatic conditions, heat stress in construction workers should not be totally discounted. The likely occurrence depends on the ambient air temperature, the relative humidity, radiant heat and the amount of air movement. Thus hot humid weather, especially for people carrying out heavy manual work in confined spaces with little air movement, provides the conditions for this hazard.

Another area may be underground, during tunnelling operations where heavy manual work is being undertaken in the vicinity of heavy machinery, and where the ventilation standard is not as good as it should be. The symptoms are headache, giddiness, fainting or muscular cramps, and victims should be taken to a cool place, provided with fluid to drink and allowed to rest. Where the case is severe and the casualty appears to be ill, develops a high temperature and does not perspire, urgent medical attention is necessary.

Noise

Damage to hearing, usually irreversible, may be caused by exposure to high noise levels from plant and machinery on site or in a workshop. Sound levels are measured by meters that indicate the sound pressure levels with a weighting factor to imitate the characteristics of the human ear, quoted as dB(A). The generally acceptable sound level upper limit is 90 dB(A) where the sound is reasonably steady and exposure is continuous for eight hours in any one day. Where the sound level fluctuates or the exposure period is other than eight hours, the daily personal noise exposure may be calculated by the use of formulae given in the HSE guidance on the Noise at Work Regulations 1989 *Noise at work*.[17]

In addition to the continuous exposure limits there is a requirement that the unprotected ear should not be exposed to a peak sound pressure level exceeding 200 Pa. Noise reduction should in the first instance be made by engineering control, e.g. more efficient silencers to the exhaust systems of air-cooled diesel or petrol-engined plant and pneumatic-powered equipment, sound-reducing covers for compressors, careful and regular maintenance of plant and machinery with particular attention given to the prevention of unnecessary vibration, and the use of acoustic screening of noisy fixed equipment where possible. Ear protectors should be a last resort and when they are used they must be carefully selected to suit the noise levels expected. Employers must provide protectors to all workers likely to be exposed to 90 dB(A) or 200 Pa and ensure that they are used — employees must wear them. Furthermore, protectors must be provided to employees who are exposed to between 85 and 90 dB(A) if they ask for them.

Vibration

A most common injury is caused by continued exposure of the hands to high frequencies of vibration from tools such as pneumatic hammers, concrete breakers, drills and chipping hammers. It is usually known as vibration white finger. It starts with a slight tingling or numbness in the fingers and eventually causes whiteness to the tips. Attacks may last about an hour and end with a sudden rush of blood to the affected tips, often causing considerable pain.

This occurrence can be reduced by the use of vibration isolators, by the operator exerting the lightest pressure on the tool commensurate with proper control and the provision of breaks from the work on to non-vibration work for the operator. The maintenance of good circulation and warm hands is also helpful and the wearing of comfortable gloves can be an asset.

Ionising radiation

Sealed radioactive sources are widely used in the industry to check welded joints in pipelines, for example, and for other non-destructive testing. Experience has shown that site radiographers are more likely to receive serious overdoses of radiation than are their factory-based counterparts. Those working in nuclear power stations, processing plants and laboratories may also be exposed to ionising radiation. The effects of exposure to ionising radiation may be radiological dermatitis, skin burns, loss of hair and bone cancer. Compliance with the Ionising Radiations Regulations 1985 together with the HSE's Approved code of practice *The protection of persons against ionising radiations arising from any work activity*[18] is, of course, essential, and this should obviate the likelihood of persons receiving excessive exposure.

In brief, only classified persons are allowed to undertake site radiography and their work must be carried out under the supervision of an appointed radiation protection adviser who should ensure that the work is carried out in accordance with the requirements of the regulations and that any local rules relevant to the work are observed.

Other workers (not involved with the radiography) and even the public may be at risk from scattered radiation. This may be reduced by the work being carried out within a suitable enclosure from which all but the classified workers are excluded. Where this is not reasonably practicable a designated area must be established, the boundary of which should be a safe distance from the exposed isotope. The boundary must be clearly indicated with warning signs or lamps at appropriate intervals and consideration should be given to posting sentries along the boundary as an extra precaution to ensure that no unauthorised person enters the danger area.

Compressed air

Working in air at pressures above atmospheric pressure, for example on diving activities, tunnelling or caisson construction may result in compressed-air illness. In its mildest form (type I) this may take the form of skin irritation, slight pains in the joints and tightness of the chest, but with the more serious form (type II) severe pains develop in the joints, and dizziness, unconsciousness and even death may occur. A particularly distressing type of compressed-air illness is aseptic bone necrosis, which affects the ends of the long bones in the body at the shoulder, hip and knee joints, for example. It leaves the affected person with symptoms similar to arthritis, varying from extreme discomfort to total disablement.

The cause of the illness is associated with too-rapid decompression and, to reduce the likelihood of its occurrence, compliance with acceptable decompression procedure and tables is necessary. The Work in Compressed Air Special Regulations 1958 include decompression tables for use in tunnelling and caisson work, but subsequent research has identified a decompression procedure which gives better protection than the 1958 tables, as they are known. This is incorporated in decompression tables known as the 1966 Blackpool tables and the Health and Safety Executive advise their use. Since the 1958 tables are the statutory requirement, contractors wishing to use the Blackpool tables must first obtain official approval from the Health and Safety Executive. Decompression procedures in respect of diving operations are covered by the Diving Operations at Work Regulations 1981.

Lasers

Lasers are used in the construction industry as an aid to setting out works which involve straight lines as, for example, in tunnels, runways and dredging work.

Eye injury, which could be severe, may occur if someone looks directly into the beam of a laser in the direction of the source. Moreover, contact may cause skin burns. It is important to select safe instruments and where possible to ensure the separation of personnel and the beam by means of effective barriers; furthermore, notices should be displayed to warn people against looking down the beam. Where people need to approach the beam, appropriate eye protection against the particular type of laser must be provided.

Biological hazards

Biological hazards experienced in the civil engineering and construction industry are likely only to affect those engaged on waste land sites or sewer outfalls or in sewers contaminated by vermin. The disease that is sometimes caught by these workers, is leptospiral jaundice or Weil's disease. This is a feverish condition caused by ingestion of food or water contaminated with the urine of rats. Where not treated early and properly the disease can be fatal. The contaminated sludge in sewers may also enter the skin or be absorbed through mucous membranes of the eyes, nose and mouth (*see also* p. 114).

5

Safety policies

Health and safety policies
Legislation
Section 2(3) of the HSW Act states

> Except in such cases as may be prescribed, it shall be the duty of every employer to prepare and as often as may be appropriate revise a written statement of his general policy with respect to the health and safety at work of his employees and the organisation and arrangements for the time being in force for carrying out that policy, and to bring the statement and any revision of it to the notice of all his employees.

The opening phrase is covered by the Employers Health and Safety Policy Statements (Exemption) Regulations 1975, which exempt from these requirements 'any employer who carries on an undertaking in which for the time being he employs less than 5 employees'. For the purposes of this exemption, all the employees of an undertaking count whether they are employed in one or several sites or establishments.

Put another way, the HSW Act demands that employers of five persons or more

- state in writing their general policy on health and safety
- describe the organisation and arrangements for carrying out that policy
- revise the statement whenever appropriate
- bring the statement to the notice of all employees.

Definition of employer
Professional staff should not be confused by the use of the word employer in the HSW Act and Employer in the ICE Conditions of Contract. The former means 'one who employs persons in return for a wage or salary' while the latter is essentially the promoter of civil engineering works otherwise known as the Client. Under the HSW Act, employers include all authorities, undertakings and companies, consulting engineers, architects, contractors, subcontractors and anyone else who has people under a contract of employment.

Health and safety policy statements
Advice to employers
Policy statements prepared and issued by employers to comply with the HSW Act should be in terms that can be clearly understood by their

employees. The Health and Safety Commission advises employers to emphasise the legal significance of the statement by referring to the HSW Act in the opening paragraphs, and the notice should make clear not only the intentions of the employer but also the obligations of employees. Policy statements should indicate how the company is organised with respect to the health and safety responsibilities of management and indicate its commitment to providing information, training and advice to employees.

Any arrangements the company has made for the establishment of safety committees and for consultation with safety representatives as required under section 2(4), (6) and (7) of the HSW Act should be covered in the policy statement. Finally, the statement should be signed by the chief executive, the managing director or other responsible head to make clear the commitment to the policy at the very highest level.

Policy statements applied to construction work

The framing of the policy will, of course, be influenced by the type of undertaking or business conducted by the company. Most employers engaged in promoting, designing, constructing and demolishing civil engineering works and buildings have responsibilities, both to their employees, and to other people, who may be employees of other companies associated with the project, visitors or members of the general public.

The opportunity should not be overlooked to emphasise to all employees the need to act at work in such a way that the health and safety of the public and of people who work for other organisations are also protected. A health and safety policy statement should therefore contain the following basic elements

- that the company has a legal duty, as far as is reasonably practicable, to provide safe and healthy working conditions for its employees
- that the company will conduct its work in such a manner that the health and safety of others, besides that of its employees, is not adversely affected
- that employees are required to co-operate with the company in preserving their own health and that of other employees and anyone who might be affected by their activities at work
- that an organisation and chain of authority in matters of safety has been set up
- that safety information, instruction and training will be provided for all employees as and when necessary
- that the company will encourage the work of a safety committee and co-operate with safety representatives if that is the employees' wish.

Typical policy statements

Health and safety policy statements for organisations involved with construction have many similarities. Differences lie in the length and method of presentation. Some employers choose to display the statement on notice boards while others include it in a safety handbook. Either method is acceptable provided the message reaches all employees.

One consideration is the problem of the changing organisation, particularly in a large company. Because the organisation for implementing the health and safety policy should contain the names of staff with posts of responsibility, it is usual to publish an organisation chart as an appendix to the main policy statement to avoid the need to reprint and reissue it every time there is a managerial change. Some organisations publish their policy statements in a loose-leaf binder for easier updating.

Policy statement of a major employer. A good example of a health and safety policy of a major employer is that of British Rail because of the wide variety of responsibilities it has. The railways employ a large number of staff on maintenance duties, and construction work is carried out by direct labour and by independent contractors. In addition, the safety of the travelling public is of paramount importance.

In British Rail the primary responsibility for safety rests with the Chairman of the Board who issues a personal policy statement which emphasises the Board's legal duty and its commitment to health and safety and welfare of its employees and of those who may be affected by their activities. This is displayed on all staff notice boards. He also issues instructions to each of his five directors, one being the Director of Civil Engineering, to prepare and implement detailed safety policies in each of the six operating regions of British Rail. These are known as local safety policy statements and they are issued to staff from Supervisor grade upwards. It is normal practice for the General Manager of the region to sign the Regional Engineer's policy statement, ensuring an essential link in cascading the policies to supervisors.

A typical local policy statement includes

- responsibility: the names of staff responsible, together with their locations and delegated duties
- safety representative: arrangements for committees and trade union consultation on safety matters
- inspections: management inspections to identify risks arising out of unsatisfactory workplaces, working systems or behaviour
- induction: safety instruction for new employees
- training: arrangements for further training
- accident reporting: internal arrangements for reporting and recording personal injuries and for accident investigation
- plant: maintenance and operator training
- materials: approval of substances and materials that may be of a hazardous nature
- routine processes: establishment of safe working systems
- protective clothing: arrangements for selection and wearing of protective clothing and safety equipment
- fire/explosion: reference to a separate fire manual and appointment of fire wardens
- welfare: assistance with staff welfare problems
- first aid: provision of first-aid equipment and first aiders
- contractors and other non-railway personnel: reference to a separate

booklet entitled *Track safety handbook* which must be supplied to contractors and other railway staff responsible for contractors' safety
- safety advisory service: how to seek advice.

The statement concludes with a number of appendices giving local information which must be completed by the local manager concerned, and lists of employees who should receive the various other railway safety publications and booklets.

The general safety policy of British Rail is typical of most major employers. However, the local or particular policy statement, which reflects more closely the special hazards of the industry or the service it provides, varies in content depending on the emphasis needed. Nevertheless, policy statements are expected to outline the measures by which the management safeguards its employees, contractors and other people on the premises and ensures the safety of the public, who could be at risk if the procedures laid down are not complied with.

Policy statement of a contractor. A typical safety policy statement of a building or civil engineering contractor is suggested here. The fictitious contractor, Constructwell Limited, includes its policy statement in the company safety booklet.

To all Employees of Constructwell Limited
Health and Safety Policy

You will know from your Contract of Employment that you are an employee of Constructwell Limited. Under the Health and Safety at Work etc. Act 1974 the Company is required to have a Health and Safety Policy and to bring it to your attention. You are required under the same Act to follow the safety procedures adopted by the Company.

This notice sets out the Company's Health and Safety Policy and the organisation and arrangements adopted for its implementation. It sets down the procedures which management will apply in the interests of health and safety of all employees.

I wish, however, to emphasise that each and every one of us has a duty to take reasonable care of ourselves and of those who work with us. We must all work together to prevent accidents and the hardships that follow.

Signed . . . . Issued . . . . 19. .
Chairman

General Policy Statement

It is the policy of Constructwell Limited

(a) to ensure the health and safety of all its employees while at work by providing, so far as is reasonably practicable, working environments that are safe and without risk to health

(b) to conduct the undertakings in such a way as to ensure, so far as is reasonably practicable, that people not in its employment, but those who may be affected, are not exposed to a risk to their health

(c) to recognise its obligation to meet all relevant legislative requirements pertaining to health, safety and welfare which apply to any of the Company's undertakings

(d) to organise and arrange its affairs to ensure effective implementation of the policy.

Operation of the Policy

In carrying out its general policy it is the practice of Constructwell Limited

(a) to specify in writing managerial responsibility and accountability for the safety, health and welfare of its employees and for the health and safety of others who may be affected by its undertakings and to bring this to the attention of all employees

(b) to ensure that appropriate safety training and instruction is provided for all employees and that accident prevention is included in all relevant training programmes, particularly those for graduates, trainees, apprentices and other young employees

(c) by a programme of regular presentations and publicity to sustain an awareness of the need to prevent accidents and health risks in the minds of all its employees

(d) to take into account when planning its work any aspects which help to eliminate risk of injury or health damage to employees or others

(e) to require subcontractors to operate a safety policy no less stringent than that of the Company; to provide safety training for subcontractors' employees and self-employed as necessary

(f) to make appropriate accident prevention arrangements by effective liaison with employers other than subcontractors who have employees working at the same work place as the Company

(g) to encourage the discussion of health and safety matters at all levels including the setting up of arrangements for joint consultations with employees through their appointed safety representatives.

Organisation

Managers at all levels are responsible for the health and safety at work of employees reporting to them and others to whom the company has a duty. They will ensure that safe working conditions and welfare facilities are provided in accordance with statutory regulations and company policy and, so far as is reasonably practicable, carry out works in such a manner that there is no risk to other people.

The line management structure of Constructwell Limited is published from time to time and displayed on principal notice boards.

The Safety Service

A senior corporate director will be responsible for ensuring that an effective safety service is provided. He will appoint a company safety service manager and a number of suitably qualified safety advisers.

The main responsibilities of the safety service are

(a) to ensure that managers understand their duties as outlined above and to assist and advise them in the execution of such duties

(b) to inspect and report on places of work to ensure that safe systems are being employed in compliance with statutory regulations and company practices

(c) to warn managers and other employees of existing and potential health hazards

(d) to investigate accidents, analyse accident data, report and make recommendations to prevent recurrences

(e) to ensure that all reports and records required by statutory provisions are made and maintained

(f) to promote safety training for all employees and, where necessary, subcontractors' employees and self-employed.

Constructwell Limited is a sufficiently large company to warrant the establishment of a safety service with a number of safety officers. Smaller contractors have to organise their safety arrangements in other ways, as suggested later in this chapter, although in principle the general policy on health and safety would be common to most civil engineering contractors.

Policy statement of a consulting engineer or architect. The following is a suggested safety policy statement for a firm of consulting engineers or an architectural practice.

Victoria Street and Partners
Health and Safety Policy Statement

The Health and Safety at Work etc. Act 1974 (the Act) places a duty on every employer to prepare and, as often as may be appropriate, revise, a written statement of his general policy with respect to the health and safety at work of his employees and the organisation and arrangements for carrying out that policy, and to bring the statement to the notice of all his employees.

The Company Policy

It is Victoria Street & Partners Policy in so far as is reasonably practicable

1. to provide and maintain places of work for their employees in conditions that are safe and without risk to health together with adequate facilities for first aid and welfare
2. to conduct their business in an organised and responsible manner and to adopt systems of work that preserve the health of their employees and other people concerned in business activities
3. to provide lines of communication to enable employees to co-operate in promoting and developing effective measures to ensure their health and safety at work
4. to give information, instruction, training and supervision as necessary to ensure that their employees are aware of their responsibilities under the Act
5. to make available personal protective clothing and equipment for employees whose occupation or place of work is likely to be hazardous.

The Employees' Responsibilities

The Act requires that employees while at work

1. shall take reasonable care of their health and safety and that of all other persons who may be affected by their acts or omissions at work
2. shall co-operate with their employer in all safety, health and welfare matters and maintain their place of work and all their equipment in a tidy and safe condition
3. shall not interfere with or misuse anything provided in the interests of their health and welfare.

Further details of the employees' responsibilities in connection with the Act are given in the Company Safety Handbook.

The Health and Safety Organisation

The means by which Victoria Street & Partners safety policy is implemented is through line and project management. Annexe A outlines the line-reporting relationships within senior management with respect to health and safety.

The responsibility for safe systems of work on projects and for the health and safety of employees working on projects away from the company premises rests with project managers, one of whom is assigned to every project by the appropriate partner in charge.

Staff with Special Responsibilities for Safety

The partner of Victoria Street & Partners nominated in Annexe A is accountable to the senior partner for the health and safety policy. He has the direct staff assistance of

1. The Company Safety Adviser, who is responsible for advising the partners of their duties under the Act in relation to
   (a) the health and safety of employees while engaged in business duties while outside the company premises
   (b) the organisation of safe systems of work so as to protect other people from risks arising out of work activities
   (c) training and informing employees about their duties under the Act, the hazards of their work and precautions to be taken.
2. The Administration Manager, who is responsible for the health, safety and welfare of employees while on company premises.

Display of this Policy Statement

The latest revision of this statement, together with the organisation chart Annexe A, will be displayed on principal notice boards by the Administration Manager.

Signed . . . . Issued . . . . 19 . .
Senior Partner

For every project Victoria Street & Partners appoint a project manager, who is responsible for all aspects of the project to which he is assigned. His duties will be outlined within his terms of reference. Other firms of consultants prefer to use the title project engineer. Whatever title is adopted, it is essential that one person of suitable experience and seniority is responsible for each project and that the reporting relationships on that project are clearly understood by all staff working on it. On large projects the duties of the project manager may have to be delegated, and it is then necessary to give clear terms of reference to all subproject managers if the health and safety objectives are to be achieved.

It must be assumed that Victoria Street & Partners is a large firm situated in one principal office. Smaller companies may well combine the health and safety responsibilities of the company safety adviser with that of the administration manager. Other firms have several offices and the duties of the two people in our fictitious consultancy must be covered in other ways. The organisation diagram should clarify these responsibilities to the employees. The employees of Victoria Street & Partners do not have a safety committee, otherwise there would be some reference to this in the policy statement. Where employees wish to have safety committees and safety representatives, employers must co-operate with them.

Organisation

The type of business

The organisation by which a company fulfils the aims of its health and safety policy reflects its particular business purpose. Whereas contractors are concerned almost entirely with the safety of their employees and others on construction sites, the major employing authorities (e.g. the electricity generating, gas and water companies, airports authorities and British Rail) have a particular responsibility for the protection of the public. As a result their safety organisations reflect this need.

The responsibility for the safety of employees and others involved in construction is invariably that of line management. The role of the safety director and his organisation in this event is advisory. However, where design and specifications are involved, the safety director has a more positive role to play because he may also be required to take responsibility for the health and safety aspects of research, development and assessment of safe working practices.

Duties of management

The chief executive, managing director or senior partner must assume ultimate responsibility for the safety policy and he normally sets up an organisation based on the line management structure to assist him. Figures 5.1–5.4 give examples of the reporting lines of staff with managerial responsibilities in typical authorities, contractors and consulting engineers.

It is common practice in organisations of more than about 50 employees for the chief executive to appoint one of his senior colleagues as safety director to take special responsibility for safety on the management board. While the appointment of safety director is usually combined with other duties, his job is to set up and maintain an organisation to promulgate the safety policy and report to the board or management committee on all safety matters. In smaller companies the chief executive may carry out these functions himself.

The management committee must make adequate financial provision for safety, health and welfare, and one of the duties of the safety director is to manage the budget allocated to maintain the safety organisation and provide information and training. Financial provision for these also includes sufficiency of price and time in tenders and budgets for safety measures to be taken in design, construction or demolition. Estimators and project managers must be made aware of this. The duties of the safety director or the senior manager responsible to the chief executive for safety matters may be summed up in typical terms of reference which are given later.

The diverse nature of the construction industry puts considerable onus for safety awareness on the managers of departments, subsidiary offices, project offices and sites. It is probably true to say that the success or failure of a safety policy rests with this level of management. One of the important tasks of a safety director is to ensure that all construction and site managers and other senior staff are totally committed to the company safety policy. Terms of reference or standing instructions must be provided for senior staff and the safety director will need to meet senior staff from time to time to ensure that their duties are understood and carried out.

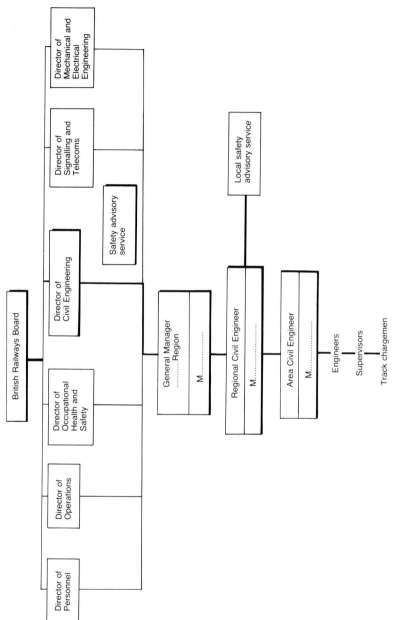

Fig. 5.1. *A major employer — British Rail: structure of responsibility for health and safety in civil engineering*

SAFETY POLICIES 149

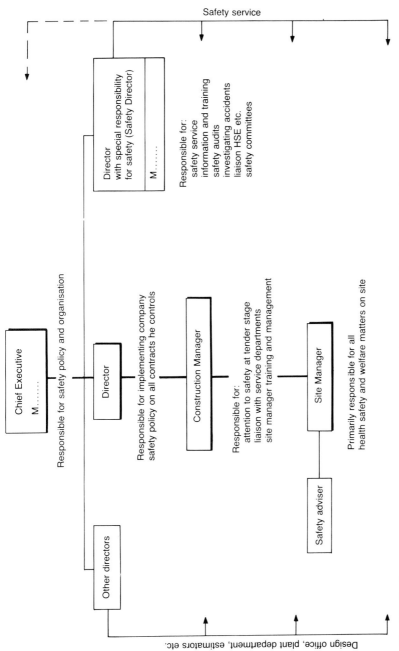

Fig. 5.2. Civil engineering contractor: company structure of responsibility for health and safety

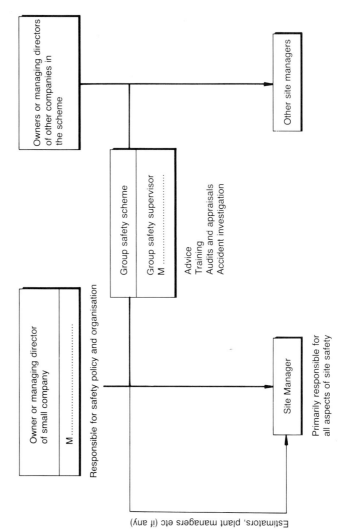

Fig. 5.3. Small contractor in group safety scheme: structure of responsibility for health and safety

SAFETY POLICIES 151

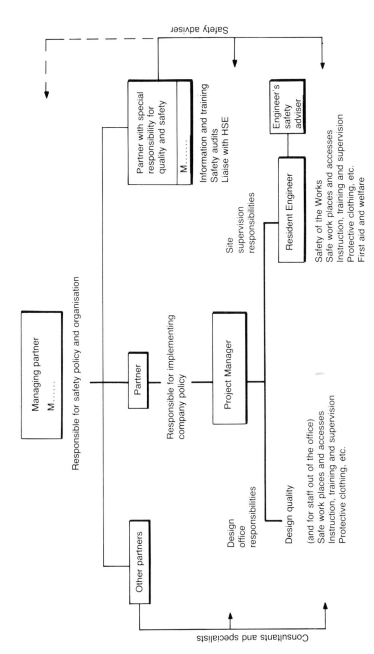

Fig. 5.4. Consulting engineers: structure of responsibility for health and safety

Safety organisation of a major employer

Figure 5.1 shows the structure of responsibility for health and safety within the civil engineering departments of British Rail. The responsibility for implementing the Board's safety policy rests with line management from the track chargeman up to the Chairman. The safety departments set up by the Director of Civil Engineering and the Regional Civil Engineers are clearly labelled advisory to emphasise this point.

Copies of the full local safety policy statement are given to all staff down to Engineer level and, in a modified form, to Supervisor level. All policy statements leave blanks for the names of staff to be inserted, leaving no doubt about either the identity of the person who is to carry out the actions or the identity of the person to whom he or she reports. The safety departments themselves have organisational structures and reporting procedures to suit the measure of hazard that the particular undertaking poses to workforce or public.

Safety organisation of a contractor

For a contractor the primary objective of the safety organisation is the prevention of accidents on construction sites and the diagram has to show the line management responsible for achieving this. Figure 5.2 shows the structure of responsibility for health and safety in a major building and civil engineering contracting company. In this case, all the directors are responsible to the Chief Executive for the safety of the sites under their jurisdiction and that responsibility is carried down through, but not delegated to, the construction managers and site managers.

One of the directors has special responsibility for providing the safety service as others have responsibilities for the estimators, design office and plant departments. All these service departments contribute to the corporate safety policy in roles that essentially support the construction and site managers.

Terms of reference for the safety director. Typical terms of reference for a contractor's safety director are as follows.

> To ensure so far as is reasonably practicable that the company meets its legal obligations with respect to health, safety and welfare of its employees and of others who may be affected by its acts or omissions. All are set out in the company health and safety policy statement.
>
> In particular the safety director must
>
> (a) administer a safety organisation, appoint appropriate staff and provide them with terms of reference
> (b) ensure appropriate information, instruction and training for all staff and operatives
> (c) provide facilities for first aid and welfare
> (d) prepare annual budget estimates for safety, health and welfare facilities and manage the allocation of funds accordingly
> (e) be responsible for reporting injuries, diseases and dangerous occurrences and keep the Board regularly informed of such events
> (f) ensure adequate provision of time and costs of safety measures in all design, construction and related activities

(g) ensure that any safety committees proposed by staff or operatives are properly constituted, attended by management at the appropriate level, accurate records of meetings are made and kept, and that the resolutions put forward by safety representatives are acted on to the committees' satisfaction.

Terms of reference for a construction manager. The terms of reference for a construction manager responsible for a number of sites each under the control of a site manager are primarily to see that company policy is implemented on those sites. Construction managers must ensure that adequate consideration has been given to site health and safety at tender stage and that the safety department, plant department, design office and any other support facilities are made available to the site manager.

Construction managers must also see that the site managers have been thoroughly trained in their duties for health, safety and welfare matters and that the duties are carried out. Advice given by safety advisers must be implemented and, where advice is ignored by any site manager, the construction manager must take appropriate action. Through liaison with the company safety adviser, the construction manager will ensure that the site managers are given all the support they need.

Safety responsibilities of a site manager. Typical safety responsibilities of a site manager are

To organise the site in such a manner that it is safe both for company employees and other persons who may be affected.
In particular the site manager shall

1. comply with the HSW Act, the Construction Regulations and other relevant legislation
2. appoint the necessary staff to carry responsibility for site safety and welfare and to provide them with the authority to act
3. complete the statutory notices and liaise with HM Factory Inspector and the chief officer of the local fire brigade and ambulance service
4. provide time and funds for safety measures
5. ensure that all staff and operatives have the necessary training and follow correct working procedures
6. provide protective clothing and safety equipment where necessary and see that it is used
7. arrange for the company safety adviser to set up a site appraisal
8. ensure that the procedures to be adopted in the event of an accident are clearly understood
9. make arrangements to supervise the safety of subcontractors' work and self-employed staff and labour.

Safety responsibilities of the safety adviser or officer. Under the Construction (General Provisions) Regulations 1961, employers of more than 20 workmen, whether they are employed on one site or several sites, must appoint a qualified person, or persons, to advise on safety and to supervise safe conduct of the work.

In the construction industry the function of the safety adviser may be as a full-time head of department in a large company or authority or a part-time specialist with a small firm, and there are many variations between the two extremes. Some smaller contractors operate group schemes that share the services of a full-time safety professional.

Safety advisers employed by contractors should be thoroughly experienced in construction techniques and legislation and preferably qualified by membership of the Institution of Occupational Safety and Health. The safety adviser reports to the safety director, who provides his terms of reference. He may well have a number of staff to assist him — indeed in some larger firms his title is Director of the Safety Service, and he has several safety advisers reporting to him.

The position of safety adviser is essentially an advisory one. Hence the preferred title of safety adviser rather than safety officer. *No action of the safety adviser must diminish the clear authority of line managers in matters of safety.* Only in exceptional circumstances where health is at risk or there is imminent risk of an accident should a safety adviser exercise the authority given him by the Directors to override the processes of line management.

The function of a safety adviser with a firm of contractors is

- to advise management on its duties under health, safety and welfare legislation and any revisions to it
- to report on the fulfilment of the objectives of the company's safety policy
- to advise on safe working practices, safety of plant and equipment, environment health problems, protective clothing and permit-to-work procedures
- to prepare and circulate safety memoranda, guidance notes, posters and other printed information
- to organise safety training courses for staff and operatives
- to carry out site safety, health and welfare audits
- to take responsibility for the company accident record book, investigate and report on all accidents and recommend measures to prevent recurrence
- to advise on the special problem of subcontractors and self-employed operatives on the company sites
- to attend safety committees as required and give advice to managers in their consultations with Safety Representatives
- to act as the focus in the company's correspondence with HM Factory Inspectorate and professional bodies specialising in safety and accident prevention.

Management contracting

Compared with ordinary contracting, management contracting, the process whereby the management contractor assigns the majority of the work to a number of works contractors, requires a different approach for the organisation of site safety. Whereas every works contractor is responsible for the safety of his own employees, it is clearly necessary for firms to co-ordinate their activities so that they act responsibly to one another and as an integral part of the total project. The need for co-ordination is emphasised by the fact that many subcontractors employed by the works contractors have fewer than five employees and are therefore exempt from preparing policy statements.

It is essential that the conditions of contract require the Management Contractor to take full responsibility for site safety, including complying with all the general site legal requirements and to co-ordinate the safety practices of all companies contributing to the project. It is strongly recommended that promoters, clients or Employers (to use ICE Conditions terminology) require to see and approve the safety policy statement of the management contractor and his proposals for the safety organisation on the project before the contract is signed. Attention is drawn to the HSE CONIAC publication, *Managing health and safety in construction: management contracting* (see Appendix 4).

Organisation of health and safety policy for management contracting. A typical statement of the organisation for implementing the health and safety policy of a management contractor is as follows. It will be noted that the Constructwell management contracting company is part of the Constructwell Group.

Constructwell Management Contracting Limited
Organisation of the Safety Policy

General Considerations

The Company recognises that the safety and health in all those affected by its undertakings is an inherent part of its total business and that all staff can make significant contributions to the prevention of accidents and ill health arising from its operations.

The health and safety policy of the Constructwell Group is applicable to its Management Contracting operations and the statement shall be brought to the attention of the staff and operatives of subcontractors in addition to Company staff.

The means by which the Company implements its health and safety policy is through line management. The primary responsibility for safety rests with the Project Managers, one of whom is appointed for every project. The Project Managers are to be supported in every respect by the directors and all other managers and staff. In addition, assistance is provided from the various service departments as follows.

Duties of Service Departments

The Constructwell Group Safety Service is available

1. for advice on all matters of health, safety and welfare
2. to carry out safety appraisals, audits and inspection
3. to investigate accidents
4. to promote safety training.

The Proposals Department must agree with Clients and their professional advisers the collective and individual responsibilities for site safety and in particular the arrangements for common services.

The Planning Department will

(a) ensure that tender and works programmes make due allowance for the safety aspects of the work

(b) bring to the attention of the Project Manager any areas of high risk for discussion with the Safety Service.

The Purchasing Department

(a) will ensure that contract documents encourage safe working and make proper provision for subsequent payment
(b) must study subcontract submissions to identify any factors that could lead to unsafe working
(c) must advise the Project Manager of any special or unusual requirements in the contract documents that might affect the safety of the workforce or others
(d) must ensure that provision for safety requirements is made in the project cost plan.

The Central Plant Engineering Department will

(a) advise on special lifts, installation, maintenance and inspection and use of cranes, hoists, lifting gear and all other lifting appliances
(b) supervise any plant items provided by subcontractors and advise the Project Manager of any non-compliance with relevant statutory requirements.

The Design Department will

(a) carry out the design of temporary works or check and approve the design by others as requested by the Project Manager
(b) carry out, or commission and approve the design of permanent works as requested by the Project Manager; due consideration is to be made of all aspects of safety and health matters in construction, operation, maintenance and demolition
(c) note that a quality plan is to be prepared and followed for every design project.

Duties of the Project Manager

The Project Manager will give careful consideration to all aspects of health and safety on site. He will ensure that work plans of each of the works contractors include detailed provision for safety and also ensure that any shortfall in the responsibilities of subcontractors is rectified by instructions in writing.

Particular consideration must be given to common services on the site and the allocation of responsibility for their provision and maintenance in accordance with appropriate legislation, e.g.

(a) liaison with fire, ambulance and rescue services
(b) liaison with the Factory Inspectorate
(c) first-aid and nursing facilities
(d) fire precautions
(e) health and safety posters, notices, information sheets and safety training
(f) protective clothing and safety and rescue equipment
(g) site offices
(h) canteen and toilet facilities
(i) access to the works including scaffolding
(j) lifting appliances
(k) temporary lighting
(l) fuel and power supplies
(m) overhead and underground cables.

The Project Manager will arrange for the subject of site safety to be included on the agenda at every regular progress meeting and will ensure that any matters raised are dealt with. He will encourage the formation of a site safety committee comprising representatives from all subcontractors. The chair of such committees will be taken by the project manager or his nominee, a Company staff member. Where employees wish to elect site safety representatives the appropriate arrangements will be made in accordance with the HSW Act section 2(4), (5) and (6).

The Project Manager will seek the advice and assistance of the service departments on any safety and health matters outside his certain knowledge. On acceptance of the bid he will initiate a safety appraisal to ascertain any particular hazards in the construction or demolition procedures and ensure that all concerned are made aware of the risks and precautions to be taken.

Subcontracting

The practice of extensive subcontracting has been adopted in recent years in some parts of Britain. Whereas a normal civil engineering contract under ICE Conditions recognises that some of the works may be sublet with the written consent of the Engineer, many contracts are now executed largely by subcontractors, the Contractor having only a cadre of managerial staff and key operatives on site.

With regard to health and safety, this situation is akin to management contracting and the principles of demarcation and responsibility for site safety are the same. However, in practice the effective control of site safety practices is difficult to enforce when a number of small subcontractors, especially those with fewer than five employees, are engaged on one site.

Nevertheless, it is desirable that Engineers take into account the safety policy and method statements of the subcontractors when giving written approval to their employment together with the contractor's proposals for maintaining overall effective control. Attention is drawn to the Construction Industry Advisory Committee booklet entitled *Managing health and safety in construction: principles and application to main contractor/ subcontractor projects* (*see* p. 168 and Appendix 4).

Safety groups and group safety schemes

Group safety schemes are operated in a number of areas and are intended for firms whose business or turnover does not warrant the employment of a full-time safety adviser or safety consultant. To comply with the supervision requirements of the Construction (General Provisions) Regulations 1961, firms forming the group each contribute to the services of a group safety supervisor whose duties are

- to advise member firms of the requirements of the HSW Act and other relevant legislation
- to advise member firms on the preparation, organisation and implementation of their safety policies
- to organise safety information and training
- to carry out safety audits, site inspection and reports
- to carry out accident investigation and advise on action to be taken in the event of an accident

- to liaise with HM Factory Inspectorate and local authorities on behalf of member firms and the group.

Safety supervisors have to be experienced in the work being carried out and be suitably qualified.

The structure of responsibility for health and safety for a small contractor in a group safety scheme is shown in Fig. 5.3. Details on the formation of group safety schemes and suggested financial arrangements may be obtained from the Building Employers Confederation in London. While this implies that group schemes are applicable only to builders and contractors this is not so. The formation of safety groups among firms of consulting engineers or architects to carry out similar objectives but with a bias towards design and supervision is recommended.

Safety representatives and safety committees

Section 2(4) of the HSW Act provides for the appointment in prescribed cases by recommended trade unions of safety representatives from among the employees. In the construction industry this applies to members of one of the unions which is party to the Civil Engineering Construction Conciliation Board or, in building, the National Joint Council for the Building Industry Administrative Council. Where requested by two or more safety representatives, employers have a duty to establish a safety committee with the function of keeping under review the measures taken to ensure the health and safety at work of employees and others.

The Safety Representatives and Safety Committees Regulations 1977 describe the appointment and functions of safety representatives and the establishment of safety committees. The Health and Safety Commission booklet, *Safety representatives and safety committees* (*see* Appendix 4), sets out the regulations, the associated code of practice and guidance notes. The Commission makes it clear that, while employers have a duty to make arrangements with trade unions for the functioning of safety representatives and safety committees there is nothing to prevent employers and employees from making arrangements for joint consultation over health and safety at work that do not follow the provisions and advice contained in the booklet provided they are satisfactory to both sides.

The main functions of safety representatives are

- to investigate potential hazards and examine the causes of accidents at the workplace
- to investigate complaints by employees on matters of health and safety
- to make representations to the employer
- to represent the employees in consultation with inspectors of the Health and Safety Executive.

Under the strict requirements of the HSW Act, safety committees are set up as a result of trades union action to appoint the safety representatives who request the formation of the committee. However, the formation of ad hoc committees on construction sites where there are no elected safety representatives must be encouraged. Experience has shown that sites with good employer/employee consultation and good contractor/engineer relationships are also safe sites.

Safety organisation of a consulting engineer or architect

The duties of a consulting engineer or architect with regard to safety resemble those of the major employer and the contractor. On the one hand, professional staff, particularly when on assignment out of the office, are at risk of injury or health damage and, on the other hand, their business comprises the design and specification of items that might be hazardous to those who build, operate or maintain.

The safety of company staff is the responsibility of line management and the safety organisation chart should make this clear. A typical safety organisation for a firm of consulting engineers is given in Fig. 5.4. The project manager who carries the primary responsibility for the safety of staff on the project and for the quality of the design is supported by a safety adviser who acts in an advisory capacity. The considerable diversity of design work with which a consulting engineer is involved makes it impracticable for the safety adviser to advise effectively on all matters of design and operational safety. The project manager must therefore rely on the advice of consultants and specialists from within and outside the firm when he is called on to design items outside his common knowledge. An effective quality assurance policy will ensure that the safety aspects of designs and specifications have been properly considered.

Terms of reference for the project manager. The terms of reference in regard to health and safety issued to project managers are to ensure that the project is designed and supervised in a manner that is safe for the company employees, contractors, personnel and members of the general public. In particular, the project manager must

- organise the design to minimise risk of error and ensure that the project can be safely built, operated and maintained
- ensure that company employees both in and out of the office have safe workplaces and means of access, that they are warned of hazards and have received appropriate instruction and training, that they are adequately supervised and that they have any necessary protective clothing and safety equipment
- understand and comply with first-aid and accident reporting procedures in respect of employees out of the office on business duties
- in the case of site supervision, ensure that resident engineers have suitable terms of reference and are properly trained in health and safety matters and that the requirements of the design are clearly understood.

Terms of reference for the safety adviser. Typical terms of reference for the consulting engineer's or architect's safety adviser are as follows.

The Company Safety Adviser reports to the Partner responsible for company safety. His main function is to advise the management board on their responsibilities under the HSW Act as described in the company policy statement. He co-ordinates with the Head of Administration who is responsible for the health, safety and welfare of staff in the premises occupied by the company.

In particular he must

(a) arrange for the issue of booklets, guidance notes, memoranda, posters and literature necessary to inform staff of their duties under the Act
(b) initiate and propagate the required training in health and safety
(c) advise project managers about potentially hazardous situations likely to be encountered by staff and any other safety and health matters associated with the project
(d) in an emergency instruct the cessation of any hazardous action by a member of the company
(e) act as the focus in the company's dealings with HM Factory Inspectorate.

The Company Safety Adviser must operate and be responsible for the annual budget for expenditure on health and safety activities. He must submit to the Company Safety Director an annual report indicating

(a) the action taken and expenditure incurred on information, training, protective clothing and equipment, etc.
(b) incidents or accidents
(c) a general overview of the position of the company with respect to the fulfilment of its liability under the HSW Act.

The Company Safety Adviser is required to have attended an approved course on construction safety for management and to maintain his knowledge of safety matters at least to the level required for membership of the Institution of Occupational Safety and Health.

The duties of the Resident Engineer. The term resident engineer is applied to the senior representative of the designing or specifying organisation on a construction contract. In most cases he is the Engineer's Representative under the ICE Conditions of Contract but, in the context of health and safety, the responsibilities could apply equally to a resident architect or clerk of works.

As the senior representative of his employer on the site the resident engineer is responsible for the safety of his staff and, in addition, he is required to use his skill and judgement in preventing any errors on or omissions from the drawings giving rise to hazardous situations. He is also expected to ensure that the Employer's duties in law and his own are not compromised by the acts or omissions of the contractor.

A post entitled *Engineer's safety adviser* is indicated in Fig. 5.4. This is an appointment by the resident engineer of one of the senior and experienced members of his staff to take a special interest in safety matters at the working, as opposed to the managerial, level. This post has no particular authority under the contract or in law but it is found in practice that the nomination of a person who will, among other duties, co-operate with the contractor's site safety adviser and advise him of hazards of which he may not have been aware, brings about a good site safety discipline.

To summarise the duties of a resident engineer with respect to health and safety, he should

- ensure that working places and access to and egress from them are safe for himself and his staff
- ensure that he and his assistants have received suitable information and training in matters of safety; this includes knowledge of the HSW Act, the Construction Regulations and other legislation and codes relevant to the works
- where appropriate, appoint an engineer's safety adviser, provide suitable terms of reference and inform the contractor
- make protective clothing or safety equipment available for his staff and make sure that they wear and use it properly
- ensure that his staff, and especially those with limited experience, are properly supervised
- ensure that first-aid and welfare facilities are available for his staff
- carefully follow all instructions on the drawings or in contract documents relating to procedures that affect the safety of construction workers or anyone else
- communicate regularly with the Engineer's project manager and, in particular, discuss any details or instructions on the documents that require clarification
- administer the conditions of contract and especially the clauses relating to health, safety and welfare with due diligence; bring to the contractor's attention in writing any hazardous or potentially hazardous situation that could affect the health of anyone on or near the site; discuss site safety at regular progress meetings and record items for action.

6

Management systems for safe construction

To satisfy his legal duty to provide for the safety of employees and others from risks arising out of work activities, it is necessary for every employer concerned with construction (client, professional adviser or contractor) to recognise the hazards and manage his operations to eliminate them as far as he reasonably can.

To achieve construction work that is safe to build, operate, maintain and demolish requires, in part, a responsible attitude by clients towards the selection of their professional advisers (consulting engineers, architects, quantity surveyors, etc.) and contractors, an acknowledgement of safety requirements in the preparation of design specifications, properly organised pre-contract investigations by engineers, the recognition of construction safety by designers and the systematic central management of safe methods by contractors.

Legislation to strengthen the management and control of construction sites by placing certain duties on clients and their professional advisers as well as on contractors was in preparation in 1990. The regulations will probably be known as the Construction Management (and Miscellaneous Duties) Regulations. (*see* p.19).

Pre-contract activities
Initiative by clients

Safe workmanship is unlikely to stem from uninterested clients who are merely concerned with least cost and time and who give no encouragement to tenderers wishing to produce a quality job. There are good reasons why the client should take an active part in the promotion of safety on site, e.g.

- his image may be tarnished by the publicity following serious accidents
- the effects of resulting delays can seldom be fully recovered
- the legal liability may (in some circumstances) fall on them (*see* Fig. 6.1).

Selection of contractors. Clients should initiate the safe project by selecting only those contractors, and those consulting engineers or other professional advisers, with acceptable safety policies to work or act for them. This is done by calling for the policy statements and current arrangements for implementing them before signing the contracts and by making regular checks to ensure that those policies are effective. Clients

Fig. 6.1. Fatal accident due to lack of clearance between roadworks and the live carriageway. The client and the contractors were found guilty under section 3 of the HSW Act (courtesy Derby Evening Telegraph)

should also check on the safety record of tenderers by requiring records of fatalities and reportable injuries for the recent three years period and by rejecting from the list of tenderers those who fail to satisfy the scrutiny or fail to give a satisfactory explanation.

Contract documents. It is recommended that the conditions of contract or specification require the contractor to submit to the Engineer a safety method statement for the Works within three weeks of starting and give powers to the Engineer to issue instructions where he is dissatisfied. When the ICE Conditions of Contract 5th Edition are used, an additional clause will be needed to cover these requirements (*see also* p. 23).

The safety method statement will include

- details of the proposed safety organisation on the site
- the name and location of the appointed safety supervisor
- details of the proposed management of subcontractors' operations
- proposed safety arrangements for dealing with special hazards.

Prudent clients (Employers) will give their Engineers powers to scrutinise the safety record, policies and organisation of subcontractors and to refuse approval of their employment where they are not satisfied. Similarly, contractors should be given the right to refuse nominated subcontractors on the same grounds. It is also important that the contractor is provided with any information about the state of the land or condition of existing buildings, etc. that might affect the health or safety of workmen or anyone else on or near the site.

Sometimes the potential for loss of life or production as a result of an accident is so great that the client needs to exercise a degree of direct authority in safety matters. In such cases, the powers and responsibilities of those concerned must be clearly defined in the specification or as special requirements to the conditions of contract. Other special requirements may be necessary where statutory authorities etc. are affected by the Works.

Special requirements

The statutory authorities and other public and private undertakings that carry a special responsibility for public service and safety, issue special requirements for use in contracts where the apparatus or installations might be damaged. The most common of these are the public utilities whose pipes and cables are especially vulnerable, as are the workmen who damage them! The special requirements are usually bound into the set of contract documents issued to all tenderers or otherwise made available to potential contractors during the negotiation stage. Compliance with the special requirements is a condition of contract and any default by the contractor is subject to the appropriate action by the Engineer. The special requirements comprise a set of rules that

- describe the hazards
- tell contractors what, and what not, to do to avoid damage
- explain how to get the service cut off or temporarily suspended
- give rules for protection, marking, watching and supervision
- say what to do and who is the person to contact in the event of an emergency

Standard special requirements for the British Isles are issued by the British Railways Board, the Central Electricity Generating Board, area electricity boards, the British Gas Corporation, British Telecommunications, area water authorities in relation to water courses, water supply and foul drainage, and the British Coal Corporation or the successors of these authorities following privatisation. Other special requirements are prepared or made available to compilers of contract documents whenever work is to be carried out on or near places of particular hazard such as airports and industrial plants.

Survey and investigation

It is frequently necessary for surveyors and office-based engineers and architects to carry out topographic, geotechnical, hydrographic, structural or traffic surveys, measure work or to study conditions under which the work will take place. Many of these site investigation assignments are hazardous and deaths and injuries have occurred to staff making them.

Every project must have its project manager or project engineer whose duties include seeing that staff on assignment out of the office have safe workplaces and means of access to them (*see* Fig. 6.2). The project manager should make sure that those staff have received the appropriate instructions and intraining concerning the risks and hazards they are likely to encounter and in the ways of avoiding accidents. It should also ensure that they are provided with, and use, any protective clothing and equipment necessary.

The project manager's task is less onerous in companies that provide a good back-up safety organisation. Staff who have received basic safety training should be aware of the causes of accidents and how to avoid them. They should also be aware of the circumstances when protective clothing and safety equipment are required and where it may be obtained. The advantages of giving basic safety training to all staff as a matter of routine cannot be overemphasised.

No individual should carry out an inspection, investigation or survey unaccompanied. In case there should be an accident another person should always be at hand to call for assistance. Engineers on site assignments should leave a message at their normal place of work indicating their whereabouts and expected time of return.

Management should consider the use of portable telephones by staff on survey, investigation or inspection assignments where other communications are unlikely to be available. This applies to geotechnical, hydrographic and traffic surveys and, for example, to the survey of unoccupied buildings.

Staff must always wear stout and sensible shoes or work boots as appropriate and, unless exempt by the Construction (Head Protection) Regulations, safety helmets. Other protective equipment may be needed depending on the assignment (*see* chapter 10).

Special care should be exercised when staff need to work at a height. It is often the case that office-based staff are not trained in fall prevention and safety equipment is not readily available for inspections which have to be carried out at height. However, staff should not be put at risk, nor should they themselves take risks, merely because the time or cost of working safely appears to be out of proportion to the perceived dangers, or their duration.

Other activities where office-based staff may be at special risk and where training and supervision are essential include

- surveys of buildings for demolition
- work adjacent to, or over water
- work on contaminated sites
- work on live highways
- entry into places with excessive noise, dust or fumes
- sites where permits to work are required (*see* p. 184).

Staff who habitually carry out surveys and inspections of sites and structures are recommended to undergo basic first-aid training and to carry first-aid equipment with them (*see* p. 222). The cost and time involved in first-aid training is negligible compared with the benefits in the event of accidents, which, as we have seen, are not uncommon.

166 CONSTRUCTION SAFETY HANDBOOK

Fig. 6.2. Hydraulic scissor platform in use for bridge inspection (courtesy John Rusling Ltd)

Design and specification

The person responsible for the design of building or civil engineering works is normally one of a team of engineers, architects, graduates and technicians, each of whom is dependent on the other for producing safe, reliable and economic work, often to a predetermined budget. It is frequently assumed that, provided an engineer is qualified, his design will contain no mistakes and the work will be safe to build and operate.

In fact there are so many hazards, ranging from human error in calculations to lack of understanding of construction procedures and ignorance of science of the environment, that one engineer on his own is statistically unlikely to be able to carry out a design totally free from error. The preparation of designs, specifications and reports must, therefore, be properly managed so that the necessary level of expertise and experience is applied to all projects and so that mistakes are eliminated by an adequate system of checking.

Extent of legal and moral duties

The designer has both legal and moral responsibilities for the safety of the workforce and others. The moral responsibility of a chartered civil engineer is covered by the Institution of Civil Engineers' *Rules for profes-*

MANAGEMENT SYSTEMS FOR SAFE CONSTRUCTION 167

sional conduct: 'A member in his responsibility to his employer and to the profession shall have full regard to the public interest, particularly in matters of health and safety'.

The designer's criminal law duties are generally covered by section 3 of the HSW Act although there are circumstances where section 2 could apply. Where he is designing an article for use at work (e.g. a lifting device) section 6 could also apply. More specific regulations which cover, among other things, the duties of designers are under consideration by the HSE (*see* p. 19). A designer would be liable in a civil law action for damages suffered by reason of a faulty or incomplete design, specification or instruction. To what extent a designer would be liable for site accidents resulting from designs that are difficult to build safely would depend on the circumstances, but there is no doubt that every designer has a duty to consider the risks that will have to be taken by those building to his designs and specifications.

Design aids for safe construction

The number of ways in which a designer can contribute to safe construction are too numerous to cover here. Nevertheless, an understanding of the causes of site accidents by designers will help to focus their attention on construction details which are potentially dangerous. Since a substantial number of fatalities and injuries result from falls it is important that designers should reduce the need for men to work at a height to the

Fig. 6.3. Frameworks preassembled at ground level showing carefully controlled tandem lift and reducing the need for men to work at height (courtesy HSE)

absolute minimum. This applies to construction workers and also to operational and maintenance staff. By reducing the number of men working at a height, for example, by maximising pre-assembly of frames at ground level (Fig. 6.3), a contribution is also made to reducing the number of falling pieces of equipment and materials, which is another major cause of injury and death.

It is likely that designers will sometimes face a dilemma where the safest method of construction is not the cheapest. Their responsibilities to the client and to the contractor's workmen will not always be in accord. Items of an exceptional nature where the cheapest method of construction carries risks must be discussed with the client. This is another area in which clients can contribute to site safety.

The Construction Industry Advisory Committee (CONIAC) booklet published by HMSO, *Managing health and safety in construction: principles and application to main contractor/sub-contractor projects* (*see* Appendix 4) lists a number of items commonly encountered on construction sites which, if they are relevant, should be brought to the attention of contractors by the designers before work starts. They are

- ground conditions: contamination, instability, underground chambers and voids; extent of investigation
- existing services: underground services where known; correspondence with utilities
- temporary works: relevant information about the design of permanent works for safe design of temporary works
- temporary instability: any special features of the design which would make the incomplete structure temporarily weak or unstable
- construction materials: any health hazards
- existing plans: any available plans of structures to be demolished or incorporated in the proposed structure.

The limits of the available working space could be added to the above list (*see* Fig. 6.1). During the progress of the works, the value of good communications between the design team and the resident engineer or site manager cannot be overemphasised. Instructions on the drawings or in specifications, or sometimes the lack of instructions, can have serious consequences where they are unclear and where resulting queries are unsatisfactorily resolved. A system of communications should be set up before construction starts.

Design aids for safe maintenance

It is important for the designer of most civil engineering and building works to be thoroughly aware of the need for, and methods of their maintenance. In some cases, for example in sewer works, the requirements for maintenance are laid down by the adopting authority, giving the designer guidance on the provisions that must be made in the layout and detail design of manholes. Reference should be made to *Sewers for adoption: a design and construction guide for developers.*[1] However, for a great many large building and civil engineering works, provision for their safe maintenance must be the subject of discussions between the designer and the eventual owner or maintaining authority, preferably before design begins.

Maintenance is a hazardous occupation. The Health and Safety Executive estimate that an average of 43% of all construction industry fatalities in the five years 1981–85 occurred during maintenance, two-thirds of which were due to roofwork and other falls (*Blackspot construction. A study of five years' fatal accidents* etc., HSE, 1988).

Promoters of construction works should provide funds for, and designers should specify safe means of access to places at a height where maintenance staff are required to work. For buildings, the Building Regulations 1985 (*see* Appendix 2) provide for the safety of those who occupy or use buildings. Stairways, ramps and guards are required to be erected to given minimum standards at places where people normally have access. However, for the majority of civil engineering and building works the provision of safe access for occasional maintenance should follow the guidance given by the Construction (Working Places) Regulations, or relevant HSE guidance notes (*see* Appendix 4).

Maintenance of civil engineering works is as diverse a subject as civil engineering itself but a few examples will show how a designer and his client, owner or operator can assist in the reduction of maintenance hazards.

Choice of materials. Where the building material is selected with maintenance in mind the need for maintenance can be reduced, in some cases to negligible proportions. The repainting of a steel highway bridge, for example, is not only hazardous in itself but also lane closures impair traffic safety.

The need for maintenance can be reduced by the use of reinforced or prestressed concrete. The soffits and sides of concrete bridge spans need (or should need) negligible maintenance. However, a maintenance-free structure is not always the cheapest and the designer should present the options to his client.

Shape of structures. Where ledges, cavities, joints, internal angles and other corrosion-inducing features and dirt traps are eliminated, the need for cleaning and repainting is reduced. Box members require less maintenance than plate girders and lattice frames and, in addition, the hazards to maintenance men are more easily eliminated.

There are many other similar instances where the designer can reduce the need for maintenance and the hazards of carrying it out by careful attention to the exposed outline of the structure and to the detailing of its parts. Expansion joints, for example, give considerable maintenance problems, especially in bridges, and should be avoided at places of difficult access. Again, the designer's proposals may not always be the cheapest initial solution and discussions with the client are necessary.

Corrosion protection. The choice of corrosion protection system influences the maintenance interval. Generally the selection of corrosion protection is based entirely on economics but the designer should ensure that the cost of safety provision for maintenance workers is taken into account.

Fig. 6.4. Maintenance workers on roof using taut wire and Latchway harness system for fall prevention (courtesy Latchway Ltd)

Provision for safe maintenance. Maintenance hazards can be considerably reduced in major structures by providing safe access by means of permanent purpose-made power-operated cradles, passenger and goods hoists, platforms and gantries. These are commonly provided for major bridges, large roofs, chimneys, silos and large buildings. Where permanent maintenance platforms are impracticable or where they fail to provide complete protection the designer must consider the safe working of maintenance crews by specifying ladders, step irons, handholds, anchorage points for use with safety harnesses and fixing eyes (Fig. 6.4). He must consider how ladders, tower scaffolds, man-riding hoists, etc. may be employed.

The Joint Advisory Committee on Safety and Health in the Construction Industry drew attention in 1973 to the need for designers of buildings to provide means for erecting and tying scaffolding for future maintenance purposes. Even for existing structures not initially provided with tying positions for scaffolding, consideration should be given to permanent anchors drilled in or otherwise secured during the first maintenance programme.

The safety of roof workers deserves special mention since they suffer the highest injury rate of all maintenance staff. Wherever possible, access to

Fig. 6.5. Good tied ladder access to secured crawling board for roof maintenance (courtesy HSE)

roof areas should be by permanent, fixed ladders fitted with hoop guards but, where permanent ladders are impracticable, anchorage points for portable ladders should be provided to enable them to be securely tied (Fig. 6.5). Permanent walkways with guard rails and toe boards should always be provided wherever regular maintenance is necessary at roof level. For work on fragile roofs there are advantages if the permanent walkways are installed above the fragile covering to enable maintenance to be carried out to roof lights, gutters and ventilators. On multi-bay structures, it is sensible to provide access over the ridges. Invariably, access will also be required beyond the walkways and, wherever this is the case, permanent attachments for safety lines should be provided.

Safe access should also be designed for the maintenance of the structure and items in the roof space, especially above suspended ceilings in industrial and commercial buildings. Where permanent ladders and walkways are not viable, provision for securing temporary ladders and for anchoring temporary walkways and staging should be made. The provision of suitable eyes to which safety lines can be attached should also be considered so that when work is undertaken at heights, workmen can operate in a safe manner.

Design aids for safe demolition

For the planner of demolition activities, the availability of accurate as-constructed drawings is vital. On completion of every project, the designer should issue to the client, owner and operating authority, copies of the construction drawings modified by site revisions, giving exact details of the works as constructed. The original as-constructed drawings should be retained indefinitely by the design organisation, preferably on microfilm in safe storage, and properly archived. When modifications to existing structures are required, when total demolition is comtemplated or where there is evidence of failure or distress, the availability of accurate as-constructed drawings and calculations will reduce the risk to people engaged in surveys and demolition tasks.

There are some permanent structures that the law requires to be removed. They include offshore oil and gas installations. Other permanent structures (e.g. nuclear reactors) will have to be demolished within a foreseeable time span and their design should take into account not only the physical dangers to demolition workers but also the probable health hazards from contaminated materials.

Most building and civil engineering works, however, are designed without regard to their eventual demolition. This assumption may well be justified where provision for easier demolition increases the initial cost of the structure. Nevertheless, every project has its special circumstances and the designer is advised to discuss any possibility of partial or total demolition with the client or promoter. A study of British Standard BS 6187 *Code of practice for demolition*[2] and HSE guidance notes GS 29/1, /2, /3 and /4 (*see* Appendix 4) will give the designer advice on planning, preparation, legislation and techniques for safer demolition. Advice is given on the hazards of some special demolition work which the designer may be able to alleviate in his original design. It includes prestressed concrete chimneys, kilns, arch viaducts, jack arch floors and steel storage tanks.

Safe systems of work for design offices

The design of all building and civil engineering works should be carried out by or under the close supervision of professional staff thoroughly competent in the relevant technology. All designs must be checked and the calculations and drawings signed by the checker as well as the designer and technician. Depending on the importance of the work and on the degree of risk, the level of checking must be decided by the senior engineer responsible.

Arithmetical checking. An arithmetical check of all calculations must be made and every drawing must be checked to see that the intentions of the designer have been correctly interpreted and indicated. While the arithmetical and visual check may be carried out by a graduate, the fundamental check must be made by an experienced engineer.

Fundamental checking. A fundamental check of the design and drawings is necessary to ensure that the correct design methods have been employed, that the relevant codes of practice have been used and that the safety of those who build, operate, maintain and demolish have been taken into account. The checker must be sure that the work can be constructed safely and that any instructions indicated on the drawing, or in the specification, are both correct and unambiguous for the contractor. A secondary check

MANAGEMENT SYSTEMS FOR SAFE CONSTRUCTION 173

CONSULTING ENGINEERS
RAILWAY PROJECT QUALITY PLAN

SECTION X - DESIGN CONTROL

CLASS 1 - DESIGN CERTIFICATE

1. We certify that reasonable professional skill and care has been used in the preparation of the design of:

   with a view to securing that:-

   i) It has been designed in accordance with Railway Project Design Criteria.

   ii) It has been checked for compliance with the relevant standards in i).

   iii) It has been accurately translated into Drawings which have also been checked. The unique numbers of these Drawings are:-

   Signed..(Team Leader)

   Name   ....

   Signed..(Director)

   Name   ....

   Date   ....

2. The Certificate is accepted by The Railway Project Authority

   Signed..

   Name   ....

   Date   ....

Fig. 6.6. Example of a design certificate (courtesy W. S. Atkins Consultants)

---

of critical calculations should be made by a different method from the one employed in the original design as a further verification of the design philosophy.

Independent checking. Where civil engineering or building works whose failure in construction or use might cause multiple loss of life it is advisable (indeed it is required by authorities responsible for certain structures such as highway bridges, dams and offshore structures) for a totally independent check of the design to be made. Some authorities will allow this check to be carried out by the same organisation as that which did the original design but by a different design team. In other cases the independent check must be performed by a different company. Typical independent check certificates are indicated in Fig. 6.6.

The importance of continuing education. The importance of persistent reading about the causes of accidents and failures emanating from the design office cannot be overemphasised. This can be assisted by the circulation of journals and other literature from a library or other central source. The wisdom and experience of senior engineers retiring from the profession has to be replaced by the continuing education of younger engineers to pass on the lessons learned from the hundreds of construction site accidents and other tragedies that are reported every year. Attendance at seminars and conferences organised by the Institution of Civil Engineers and others on all health and safety matters must be encouraged.

Quality assurance in design. An effective design office system to produce safe and reliable designs and specifications is the principle of quality assurance. *Quality assurance in civil engineering* is the title of CIRIA report 109, published in 1985.[3] Quality assurance implies the application of management techniques

- to assure reliability of public safety by structured control of all stages of design and construction
- to provide positive evidence that the quality produced by the design and construction meets that specified
- to demonstrate prudent expenditure.

In civil engineering design, therefore, quality assurance goes beyond that required for reliability or safety but its adoption will certainly ensure that the works have been designed and constructed to principles that will result in their being reliable and safe.

Even without the application of quality assurance, the establishment of a project organisation with its management statement, controls and audits will ensure, as far as is practicable, that the project is error free. A design office in which a quality assurance policy has been adopted requires a quality management manual. Every project conducted to quality assurance principles requires a project quality plan.

Quality management manual. The quality management manual (QMM) is a statement of policy by the design office and defines the intent, organisation and responsibilities of the quality management functions of senior staff. The senior staff must be named individually and their reporting lines clearly stated. The QMM sets out the procedures covering all quality-related aspects of the design office work. It describes in detail the actions required to create a project quality plan and covers monitoring, records and staff training. The QMM covers all phases of design from preliminary studies to design and calculations to inspection of construction. In the QMM, names of the company quality assurance director, the quality managers and the company safety adviser are given, together with names of the chief engineers responsible for the various technologies. Fig. 6.7 shows a company structure of responsibility for quality assurance for a firm of consulting engineers.

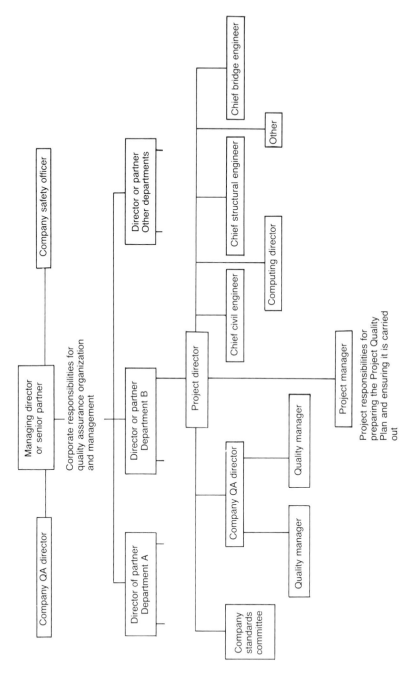

Fig. 6.7. *Consulting engineers structure of responsibility for quality assurance (Courtesy W. S. Atkins Consultants)*

Project quality plans (PQP). Every project to which quality assurance principles apply must have a unique PQP. Among other things it describes the organisation, responsibilities of staff on the project, document control, design control, quality audit and review and site supervision. These are each described in turn.

The first section of a PQP, the *organisation*, describes the purpose of the project, the companies involved and their contractual relationships to one another. It describes the management structure and identifies all the senior staff with project and technological responsibilities. Figure 6.8 illustrates the quality assurance organisation for the design of the civil engineering works for the London Docklands Light Railway Project.

The next section of the PQP is to provide a clear brief for every group or organisation involved in the project and for describing the *responsibilities of senior staff* on it. The section has to be compiled carefully so that no tasks are left undone and so that no member of the design or checking team is in any doubt of his or her reporting line.

For the proper management of a large and diverse project it is necessary to publish the *procedures for communication.* Documents include mail, telex, minutes of meetings, reports and drawings. Distribution networks and standard file references must be devised to reduce the risk of errors due to misdirected information. Figure 6.9 gives an example of the routine for dealing with incoming mail.

Design control. Design control is concerned with the engineering procedures for the technical departments engaged in the project. It lays down the

- design standards
- specification and check procedures
- methods of producing design calculations and performing design certification
- method of producing drawings and check certification
- control of changes to the design and drawings.

A typical design certificate is illustrated in Fig. 6.6.

To ensure that the requirements of the PQP have been followed and that the documentation and record systems are adequate, it is necessary to carry out a *quality audit* at regular intervals. The audit is carried out by a quality manager in the full-time presence of the project manager to an agenda that follows the audit checklist (*see* Fig. 6.10). The audit report is issued to the project director, the quality director and the project manager, who takes any necessary corrective action.

Although not strictly a design office activity, the PQP must describe the responsibilities of the *resident engineer*, if there is to be one, and clarify his reporting line with respect to the project manager and others responsible for the technical aspects of the work. Resident engineers should receive instructions on procedural matters, on health and safety aspects of the work and on their responsibility for the health and safety of staff appointed to assist them. If the construction is also to be quality assured there may be no resident engineer (see later in this chapter).

MANAGEMENT SYSTEMS FOR SAFE CONSTRUCTION 177

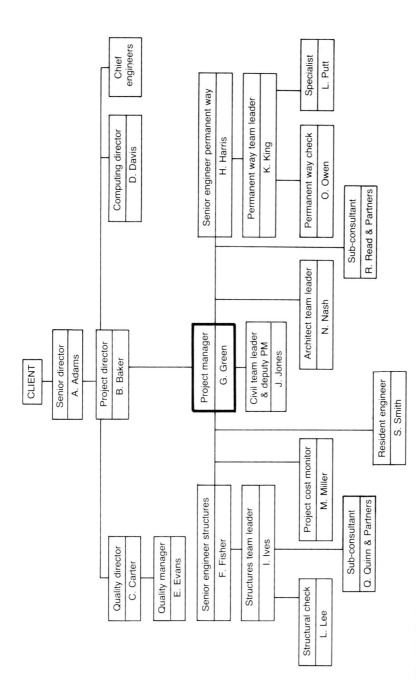

Fig. 6.8. Quality assurance organisation for the design of the civil engineering works for a railway project (courtesy GEC–Mowlem Railway Group)

178 CONSTRUCTION SAFETY HANDBOOK

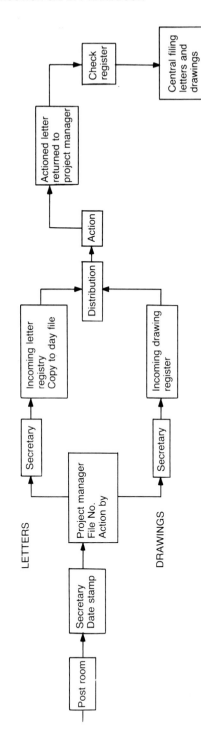

Fig. 6.9. Procedure for incoming mail

CONSULTING ENGINEERS RAILWAY PROJECT QUALITY PLAN SECTION Y — QUALITY AUDIT AND REVIEW					
A B C & PARTNERS					
London, England				Tel:
CLIENT:					
PROJECT:					
QUALITY ASSURANCE CHECK-LIST			date:	/ /8
DISCIPLINE AUDIT			Report Ref. No. AR _____	
DISCIPLINE: _____					
AUDITOR(S): _____					

Ref. No.	CHECK ITEM	YES	NO	REMARKS
1.0	STAFF			
1.1	Have all staff signed as having studied required documents?			
1.2	Are all staff approved and are CV's available in the Register of staff?			
1.3	Are staff familiar with required section of Project Quality Plan?			
1.4	Are staff familiar with the Procedures Manual?			
1.5	Are staff familiar with other relevant documents, e.g. Design, Criteria and Philosophies?			
2.0	DOCUMENTS			
2.1	Have documents issued been produced in accordance with Procedures?			
2.2	Are documents correctly filed and used, including those in other areas?			
2.3	Are documents in course of production in accordance with Procedures?			
2.4	What is programme of further document production? Is there enough time being allocated for QA?			
CHECK-LIST COMPLETED BY:		SHEET NUMBERS OF SHEETS		

Fig. 6.10. Example of an audit checklist (courtesy W. S. Atkins Consultants)

Management of construction

Safe systems of construction are achieved by the application of effective management techniques together with the facilities and motivation to fully implement the published safety policy (described in chapter 5). So that management can be sure that its construction sites are properly organised, it is useful to have a standard system for setting up a site safety organisation and for monitoring its effectiveness. This may be effected by completion of a safety appraisal for every site, followed by safety audits from time to time during construction.

Safety appraisal

The site manager is responsible for setting up the safety appraisal when the contract has been awarded. The safety appraisal is the formalised statement of responsibilities and actions required for the safe and proper execution of the contract. It includes the names of staff with managerial and functional duties and a list of the primary operations to be carried out requiring action by the staff nominated. The responsibilities of subcontractors and their senior staff must be included.

When the safety appraisal has been prepared, meetings are attended by the site manager, safety adviser, line managers and relevant subcontractors to discuss the actions, identify the hazards and allocate responsibilities in order to devise methods to eliminate or minimise the hazards in relation to the construction programme. A typical safety appraisal is shown in Fig. 6.11.

Safety audits

It will be evident that the safety appraisal is meaningless if actions required are not carried out. On sites there is pressure to ignore safety procedures to save time or to improve productivity bonus payments. It is necessary for safety audits to be carried out at intervals by the company safety adviser or supervisor. The intervals between visits to a site by the company safety adviser will depend on the degree of hazard and the experience of the site manager. The degree of hazard will in turn be influenced by the seriousness of any possible accident, especially where the public might be affected, and by the pressure of a keen construction programme. Some site managers can be relied on more than others to organise a safe site. A safety audit checklist is given in Fig. 6.12.

The audit report is circulated to relevant line managers but the responsibility for any corrective action lies with the site manager. It may be necessary for the safety adviser or even the construction manager to follow up a safety audit to ensure that any unsafe situations or practices have been eliminated.

Permits to work

Several operations are potentially so hazardous that normal training and supervision are insufficient to ensure safety. The permit to work is a document which describes the work to be done and the precautions to be taken so that the errors, omissions and misunderstandings that cause

MANAGEMENT SYSTEMS FOR SAFE CONSTRUCTION

Constructwell Limited
Safety Appraisal
Contract Concert Hall

Appointments			Appointments contd.		
Site manager		A. Adams	Registers:		
Temporary works co-ordinator		B. Baker	Scaffold		C. Carter
Temporary works inspector		C. Carter	Excavators		C. Carter
Site safety supervisor		D. Davis	Cranes		I. Ives
Safety representatives			Hoists		I. Ives
First-aid attendants		E. Evans	Fork lift trucks		C. Carter
Accident book		E. Evans	Young persons		J. Jones
Fire officer		F. Fisher	RIDDOR		C. Carter, E. Evans
Welfare facilities		G. Green	Record of trained persons		J. Jones
Form 43B		E. Evans	Permission to break live sewers		K. King
Compound		H. Harris			

Signed ..*A. Adams*..................................
Site manager
Date..

This appraisal takes effect from ....

Programme Item No.	Operation	Action req'd	By	Competent person (for register)	Comments
3	Erect tower crane	Plant dept. to provide erection details Test certificate Drivers certificate Banksmen & slingers training Test certs for S.L.I. & lifting tackle	Site manager Plant dept Safety dept " Site manager	TC driver Plant dept TC driver	Inspections F.91 C - E Inspections F.91 G PN 52 D Training req'd. Inspection F.91 J
7	Excavate pile caps & founds	All excav to be battered back or timbered	Site manager	Foreman of excav	Inspection F.91 Part 1B
8	Cut off pile tops	All ops to be issued with goggles	Site manager		Protection Eyes Regs
11	Drainage	All excav to be guarded, timbered & have safe access	Site manager	Foreman of excav	Company Practice Notes 113D and 148 D Inspections F.91 Part 1B

Fig. 6.11. Example of a safety appraisal (courtesy John Laing Construction Ltd)

accidents are minimised, even eliminated. It provides a record that all foreseeable hazards have been considered by responsible personnel.

Operations for which permits to work should be used include

- entry into confined spaces, closed vessels and vessels containing agitators or other moving parts
- work involving the breaking of pipelines or opening of plant containing steam, ammonia, chlorine, other hazardous chemicals and hot substances or vapours, gases, or liquids under pressure
- work on certain electrical systems

SAFETY AUDIT

TO .... SITE MANAGER   DATE OF VISIT ....

.... CONTRACT ....

This report follows a safety inspection of the above contract on the date stated. Items indicated by an 'X' are commented upon on the reverse side of this form. When these items have been attended to, please sign one copy of this report and return to the Safety Officer making the inspection.

1. STATUTORY DOCUMENTS						
REGULATIONS		NOTICES		RECORDS		TEST CERTIFICATES
S1 94		FORM 3		FORM 36		FORM 55
95		954		91 (Part I)		59
535		996		91 (Part II)		75
690		2345, 2347		92		80
1580		2350, 2351		2202		87
1581		2358		2346		96
OSR9		2440		BI 510		97
2003		2470				

2. SAFETY APPRAISAL	UPDATE	EFFECTIVENESS

3. SCAFFOLDS AND FALSEWORK	4. EXCAVATIONS	5. LIFTING APPLIANCES
SOLE/BASE PLATES	ACCESS/LADDERS	TOWER CRANES
STANDARDS/LEDGERS	GUARDING	MOBILE CRANES
TIES/BRACING	TIMBERING/STRUTTING	SCAFFOLD CRANES
FITTINGS	ATMOSPHERIC CHECKS	GIN WHEELS
LADDERS/ACCESS	TEMPORARY WORKS DESIGN	PASSENGER/GOODS HOIST
WORKING PLATFORMS	WEEKLY INSPECTIONS F91(B)	HOIST GATES/MESH/INTERLOCKS
GUARDRAILS/TOEBOARDS	**6. LIFTING GEAR**	OVERRUN/ARRESTER GEAR
MATERIALS LOADING TOWERS	HOOKS/SHACKLES	HOIST LEVEL INDICATOR
MOBILE SCAFFOLDS	RINGS/EYE BOLTS	RACK AND PINION HOISTS
SUSPENDED SCAFFOLDS	SLINGS	TRAINED CRANE AND HOIST
CRADLES	SPECIAL DEVICES	DRIVERS
FALSEWORK	TRAINED BANKSMAN/SLINGER	TRAINED FORKLIFT TRUCK
WEEKLY INSPECTION F91(A)	INSPECTIONS AND TESTS F91(J)	DRIVER
		SPECIAL APPLIANCES
		INSPECTIONS AND TESTS F91(C)

7. MECHANICAL PLANT	8. FIRE PRECAUTIONS	9. WELFARE
PORTABLE TOOLS	FIRE ORDERS	WASHING FACILITIES
CARTRIDGE OPERATED TOOLS	EXTINGUISHERS	SANITARY FACILITIES
WOODWORKING MACHINERY	PERSONNEL TRAINING	CANTEEN ACCOMMODATION
SITE TRANSPORT CONTROL	WARNING SIGNS	DRYING ROOM
EARTH MOVING PLANT	FIRE DRILLS	PROTECTIVE EQUIPMENT
RAIL TRANSPORT	FIRE CERTIFICATE	OFFICE ACCOMMODATION

10. ELECTRICITY	11. SITE TIDINESS	12. FIRST AID
CABLES – OVERHEAD	GENERAL TIDINESS	TRAINED ATTENDANTS
– UNDERGROUND	MATERIALS STACKING	BOXES
PORTABLE EQUIPMENT	STRIPPING OF SHUTTERING	STRETCHERS
ELECTRICAL REGISTER	SITE LIGHTING	ROOM

Fig. 6.12. (above and facing page) Example of a safety audit pro forma (courtesy John Laing Construction Ltd)

ITEM	SAFETY ADVISER'S COMMENTS

DATE.. SIGNATURE ....
Safety Adviser

THE ABOVE ITEMS HAVE RECEIVED ATTENTION

DATE.. ....
Site manager

- welding and cutting operations in areas other than fabrication shops
- work in isolated locations, locations with difficult access or at a high level
- work in the vicinity of, or requiring the use of, highly flammable explosive or toxic substances
- work which may cause atmospheric pollution
- work involving ionising radiation.

The use of permits to work is influenced by the degree of risk to which engineers, workmen, members of the public, property and product are exposed in respect of

- the type of work undertaken
- the working method used
- the location of the work
- any articles and substances used which may affect, or be affected by, the work.

The permit-to-work system requires rigorous adherence to the arrangements and in particular the authority to specified persons to issue permits to work, the signing of the permits by authorised staff and the distribution of copies of the signed permits. An example of a permit to work is given in Fig. 6.13.

Competent person

Tasks such as, for example, undertaking hazardous inspections, issuing permits to work or requiring exposure to a high level of foreseeable risk at frequent intervals should be undertaken only by specifically trained engineers or operators who appreciate the risks involved.

The expression 'competent person' occurs frequently in construction safety legislation. For example, under the Construction (General Provisions) Regulations 1961 and the Construction (Working Places) Regulations 1966 certain inspections, examinations, operations and supervisory duties must be undertaken by competent people. The onus is on the employer to decide whether people are competent to carry out these duties. An employer might do this by reference to the individual's training qualifications and experience. Broadly speaking, a competent person should have practical and theoretical knowledge as well as sufficient experience of the particular machinery, plant or procedure involved to enable him to identify defects or weaknesses during plant and machinery examinations, and to assess their importance in relation to the strength and function of that plant and machinery. He must be able to discover defects and to determine the consequences of such defects.

Competent persons are involved in many activities, including

- the supervision of demolition work
- the supervision and handling and use of explosives
- the inspection of scaffold materials before erection
- the supervision of erection of substantial alterations or additions to scaffolds, or the dismantling of scaffolds

```
SOUTHERN WATER AUTHORITY
...................... Division                                    Number ................
PERMIT TO WORK
Issue to Designated Person ......................................................................
Works and Location ..............................................................................
Work to be done .................................................................................
CONDITION CHECK (Delete where not applicable)
```

ITEM	COMMENTS	INITIALS
1. Are gas detectors and extraction fans functioning? 2. Is area free from dangerous sludge or other deposites? 3. Is machinery made safe? 4. Are electrical circuits disconnected? 5. Valves operated to make working area safe (list). 6. Can welding or cutting be carried out? 7. Are warning notices posted (list)? 8. Are water hoses, shower and eye bath working? 9. Other conditions (state).		1. 2. 3. 4. 5. 6. 7. 8. 9.

```
THIS PERMIT COMMENCES AT ......................... a.m./p.m. on ..........................................
              AND EXPIRES AT ......................... a.m./p.m. on ..........................................
SIGNED ................................. Authorised Person
```

SPECIAL SAFETY PRECAUTIONS TO BE TAKEN (Delete where not applicable)	REQUIREMENTS
1. Minimum number of men to be used. (2 must be trained in C.A.B. apparatus.) 2. 2 sets of breathing apparatus to be provided. 3. Protective clothing to be worn. 4. Detector for HYDROGEN SULPHIDE, OXYGEN DEF., METHANE, ... gas to be used prior to entry and continuously thereafter whilst men are working in the area. 5. No smoking, matches, or naked flame to be allowed in the area. 6. Arrangements for Emergency Service to be made. 7. Resuscitation apparatus and First Aid Box to be available. 8. Fire Fighting equipment to be available. 9. Any other precautions.	

```
I have read this Form and understand the work to be done and the special safety precautions to be taken. I
also understand that if the job is not completed by the expiry time, that work MUST stop until a fresh
inspection of the workplace has been carried out and a new certificate issued.
SIGNED ....................................... TIME .............a.m./p.m.  DATE .........................
```

```
TEST ON COMPLETION
On completion of the work, the following tests/observations are to be carried out : -

                                          Signed ...............................................
                                                                Authorised Person
THE ABOVE WORK AND TEST HAVE BEEN COMPLETED, THE TOOLS, TACKLE AND EQUIPMENT REMOVED AND THE PLANT/AREA MAY
GO BACK INTO SERVICE.
    Signed ....................................... Time ............... a.m./p.m. Date .........................
                    Designated Person
    SIGNATURE OF AUTHORISED PERSON ................................ Date ...................................
```

Fig. 6.13. Example of a permit to work (courtesy Southern Water Authority)

- the inspection of scaffolds every seven days and after adverse weather conditions which could affect the strength and stability of a scaffold or cause displacement of any part
- the inspection of excavations on a daily basis
- the supervision of the erection of cranes
- the testing of cranes after erection, re-erection and any removal or adjustment involving change of anchorage or ballasting
- the examination of appliances for anchorage or ballasting before crane erection.

Quality assurance in construction

CIRIA report 109, *Quality assurance in civil engineering*,[3] states that one of the reasons that quality assurance is necessary is to assure reliability of public safety by structured control of all stages of design and production. It is evident that a contract that is quality assured will satisfy the requirement of the HSW Act that safe systems have been adopted.

It is not possible to draw any correlations, let alone conclusions, from the construction sites using quality assurance compared with those not doing so. However, it is axiomatic that safety thrives in the company of efficiency, high morale, tidiness, discipline and skill. These add up to quality.

It is common practice in large construction companies to monitor the accident records reported in from various sites and to regard them as a barometer of the overall health of the job. The operation of quality assurance on construction sites has more often been applied to design-and-build contracts for the petrochemical and nuclear industries than for normal civil engineering contracts.

A typical quality plan for a civil engineering contract where quality assurance has been adopted is given in the following paragraphs. The contract was not in accordance with the ICE Conditions of Contract and there was no Engineer. The supervision was covered by the quality assurance plan and a separate organisation was appointed to monitor the quality plan.

Construction quality plan. Quality assurance applied to construction is a system that effectively and economically ensures and demonstrates that workmanship and materials conform to the specified requirements. The quality assurance programme is aimed at providing a disciplined approach to all site activities affecting quality, including verification that each task has been satisfactorily performed and that the required level of quality has been achieved.

A quality assurance programme differs from the traditional construction system, which generally relies on the experience and integrity for foremen and inspectors supported by their engineers. A quality assurance plan does not require an Engineer or an engineer's representative or his site staff. The construction manager, whose duties are outlined below, is responsible for the safe and efficient construction of the works, and an independent checker is engaged to see that the QA programme is being carried out. The duties of the construction manager include

MANAGEMENT SYSTEMS FOR SAFE CONSTRUCTION 187

- ensuring the safe and efficient construction of the works
- ensuring that the quality assurance programme is complied with.

Quality control of workmanship. To ensure that the civil engineering works are built to the specification and drawings, quality checks are carried out at strategic stages in the construction. A strategic stage is that where a significant change of circumstances occurs such as placing concrete in the completed formwork. The quality checks are carried out only by people who have been authorised in writing and each check is properly recorded before the next stage of work can proceed.

Flow sheets. Each stage of the work from start to finish of the job is identified and briefly described on flow sheets. They stipulate those

Constructwell Group
Railway Project
Date:............................

Quality Assurance Plan
Section 5 Workmanship
Revision:.....................

Page:...........................

Flow sheet
Viaducts and bridges
Fig

No.	Activity	Check	Grade of staff to check	Control Form No.
01	Excavate pile cap and clean ready for blinding	QC check	11	QA5
02	Place blinding, prepare pile tops, survey positions	Survey	13	
03	Fix shutters, reinforcement kickers	QC check	11	QA6
04	Concrete pile cap and cure	Concrete check	13	QA17
05	Fix column shutters and reinforcement	QC check	14	QA7
06	Concrete column and cure	Concrete check	13	QA17
07	Prepare column top for bearing and check position	QC check	13	QA11
08	Fix bearings and grout	QC check	14	QA11
09	Place permanent steelwork, bolt and weld	QC check	13	QA12
10	Fix deck shutters and reinforcement	QC check	14	QA20
11	Concrete deck and cure	QC check	13	QA17
12	Prepare deck for waterproofing	QC check	13	QA21
etc	etc	etc	etc	etc

Fig. 6.14. Example of a quality assurance flow sheet (courtesy GEC–Mowlem Railway Group)

Quality Inspections Group	Revn.
Railway Project	Date
	Page

Quality Assurance
Non-compliance notice No......................................

1

The workmanship/materials for the section of works
described in Section 2 below do not comply with the contract
for the reasons given in Section 3 below. Please rectify
the matter forthwith.

Signed .... Received..
        Quality Inspections Group   Constructwell Limited
Date.................. Time................... Date.................. Time.................

2 Section of Works
Reference number
Particular part of work not complying

3 Details of non-compliance

4 Rectification measures
[To be completed by contractor]

5 Because the measures set out in Section 4 above have been
taken it is agreed that the faults set out in Section 3 above
have been rectified

                                      Signed ....
                                      Quality Inspections Group

                                      Date..

Fig. 6.15. Example of a quality assurance non-compliance certificate (courtesy W. S. Atkins Consultants)

strategic stages at which quality checks are to be carried out and who is authorised to carry out the checks. An example of a flow sheet for a railway bridge excluding piling is given in Fig. 6.14. The quality control check sheets referred to are collected and distributed to the client and the independent checker.

Quality assurance review meetings. Before the start of any new form of construction or any major section of work, e.g. a bridge or a tunnel, a quality assurance review meeting is held. The meeting will be attended by the construction manager and the senior quality assurance engineer of the independent checker together with engineers and foremen responsible for the particular section of work. In some circumstances the senior designer should attend the review meeting to ensure that any construction methods required by the design are fully understood and complied with.

The agenda of the quality assurance review meeting will be

- description of the work
- construction method and sequence
- review of flow chart and quality control checks to be made
- personnel authorised to carry out the quality control checks
- design check certificates required
- any other business.

The minutes of the meetings will be signed by those present and circulated according to the prearranged list. They will form part of the quality assurance records.

Non-compliance with the contract or quality control procedure. In the normal course of events any non-compliance with the drawings or specification will be identified by the engineer responsible for carrying out the quality control checks. He will give instruction for rectification. He must then ensure that the work is properly carried out and he will not sign the control form until everything is in accordance with the contract.

When the independent checkers find that work is not in accordance with the drawings or specification or that the agreed quality control procedures have not been followed, they issue a non-compliance certificate to the contractor, who must sign that he has received it. A typical non-compliance certificate is given in Fig. 6.15.

On receipt of a non-compliance certificate, the contract must take immediate steps to identify the nature of the non-compliance. He will then

- rectify the work, or
- define the method and timing of the remedial work, or
- ascertain from the designers whether the non-compliance is acceptable and, if not, rectify the work.

When the non-compliance has been rectified, the non-compliance order is revoked by the independent checker, but if the order has been ignored, the independent checker must report the matter to the client. The client, of course, will have safeguards under the contract to enable him to apply financial pressure on the contractor if he fails to comply with the quality procedures.

Effectiveness of quality assurance

By including a description of quality assurance applied to civil engineering design and construction in this book, the authors do not wish to imply that the procedure is necessarily safer than the more usual traditional contract system. However, when applied to construction and to design systems, quality assurance methods provide a logical and orderly approach to the tasks in hand. They give experienced staff the opportunity to make their contribution to the work strategy and they ensure that the work is checked by competent people.

Civil engineering contracts employing quality assurance procedures are known to have been successful and safe, but too few have been carried out for any conclusions to be drawn on a statistical basis. Nevertheless, it can be confidently argued that some well-publicised disasters would not have occurred if their design and construction had been quality assured.

7

Safety and reliability

Risk

The imperfect nature of the world in general and of construction in particular may lead to the acceptance of a limited number of accidents as inevitable. In construction the type of activity is seldom repeated with an identical set of circumstances and it is therefore practically impossible to follow the routine systems of working or to test finished products in a way that has led other industries to achieve improved safety and reliability.

While no civil engineering or building project can be entirely risk-free, either during construction or in its completed form, there are ways in which the risks can be reduced to an acceptable level — even though that level may be hard to define. A simple example is the choice of the height of a sea wall.

Enough is known about tides to enable experts to predict the highest water level that is likely to be experienced in the vicinity in a given number of years. This is referred to as the return period. North Sea oil and gas platforms are designed for a sea state of 100 year return period.

Where a sea wall is designed for a 100 year return period, flooding, and possible loss of life, has to be accepted as a possibility once every 100 years. Politicians have to decide whether the extra expenditure on a higher sea wall to reduce the frequency of flooding and loss of life is justified. The risk can never be eradicated but it is possible in most cases to quantify the probability of failure so that decisions may be taken on the financial implications of reducing it.

Principles of risk analysis

Risk analysis is based on probabilities derived from past experience. To illustrate the principle let us use it to assess the priorities of proposals for road improvement schemes. Where for instance at a gap in the central reservation of a dual carriageway, there have been 25 serious-injury accidents in the previous five years, during which time 20 000 vehicles have passed the spot every day.

$20\,000 \times 365 \times 5 = 3 \cdot 65 \times 10^7$ vehicles have passed

The serious accident rate is then

$25/3 \cdot 65 \times 10^7 = 6 \cdot 85 \times 10^{-7}$ per vehicle per day

But from statistics collected by the Department of Transport, it is not unreasonable to expect that the accident rate can be reduced to 25% by the elimination of right turns on such a dual carriageway.

Therefore, all other things being equal, closing the gaps in the central reservation can be expected to reduce the serious accident rate by

$$0.75 \times 6.85 \times 10^{-7} = 5.14 \times 10^{-7} \text{ per vehicle per day}$$

The estimated cost of the road improvements to eliminate the right turns may then be compared with savings in the cost of the accidents. By calculating other accident blackspots in a similar way, a table of priorities for improvement schemes could be drawn up to distribute a given budget in the most cost-effective manner.

The prediction of the breakdown of a piece of equipment or, conversely, its reliability, can be expressed in the same way. From records of past breakdown, the probability that the equipment will break down in any given period may be interpolated and, with the use of subjective judgement and quantitative mathematics, an estimation of the risk of an accident to which that particular equipment breakdown played a part can be derived. Fault tree analysis and failure modes and effect analysis enable us to understand more clearly how the breakdown of one piece of equipment affects the probability of a major hazard.

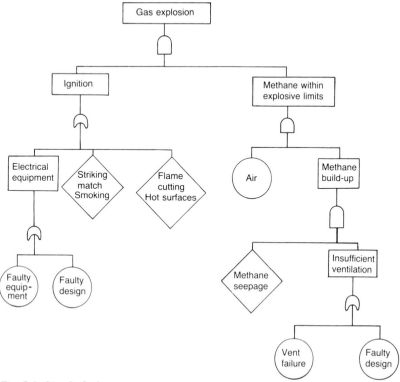

Fig. 7.1. Simple fault tree

Fault-tree analysis

A fault tree is a graphic representation of the sequence of events that could cause the hazard under consideration. Figure 7.1 is a simple fault tree. The hazard, known as the top event, in this case is an explosion in a pumping station. The explosion will only occur where there is a methane/air mixture within explosive limits AND there is ignition. Ignition can occur from electrical equipment OR someone smoking OR from hot surfaces. Similarly, electrical equipment may be faulty OR incorrectly designed. Methane within explosive limits requires oxygen AND methane build-up, which may occur from methane seepage from the soil AND insufficient ventilation.

Over the years a fairly consistent set of fault tree symbols has been established and these are shown in Fig. 7.2. The most important of the event symbols are the circle representing a basic event and the rectangle representing a fault from which further branches of the fault tree stem through the AND and OR gates.

A fault tree requires that the event is described either by success or failure. Thus, although a control valve in a pipe might seize up in any position, it must be regarded for the fault tree as being either sufficiently open or insufficiently open. Software packages are available to build up fault trees by computer graphics.

Failure mode and effect analysis

Failure mode and effect analysis (FMEA) is the study of potential failures in any part of a system to determine the probable effect on the other parts of the system and an operational success. It is a discipline that every designer goes through but with the added reliability of recording the analysis in a logical and checkable manner and of identifying the severity of the risks.

The function of FMEA are

- to identify known and potential failure modes
- to identify the causes and effects of each failure mode
- to give each identified failure mode a priority number according to

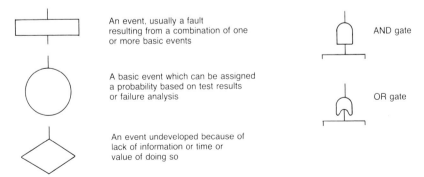

Fig. 7.2. Some fault-tree symbols

Table 7.1. Failure mode and effect analysis for support bracket of a water circulating pump

Failure mode	Cause of failure	1*	2*	3*	N†
Bracket bends	Design fault	2	5	8	80
	Construction error	4	5	8	160
Bracket corrodes	Inadequate protection	3	5	8	120
	Leak	4	5	8	160
Bracket fails	Design fault	1	8	5	40
HD bolts loosen	Incorrect torque	5	5	8	200
	Incorrect nut or bolt	4	5	8	160
HD bolts fail	Construction error	4	8	5	160
	Design fault	2	8	5	80

*1. Occurence: 1 for low probability, 10 for high probability
 2. Severity: 1 for minor nuisance, 10 for serious hazard.
 3. Chance of detection: 1 for high probability, 10 for low probability.
†N Risk priority number = column 1 × column 2 × column 3.

the probability and severity of its risk and the chance of detection before the failure occurs
- to provide for corrective action.

Table 7.1 is an example of FMEA. The risk priority number is a product of the estimates of occurrence, severity and detection, and provides a relative priority of the failure mode. The higher the number, the more serious the failure mode. The occurrence of failure is evaluated on a scale of 1 to 10, 1 indicating a very low probability of occurrence and 10 indicating a near certainty of occurrence. The severity of failure is judged on a scale of 10, 1 being a minor nuisance and 10 a very serious consequence. The detection of failure is also judged on a scale of 10, 1 indicating a high probability that failure would be detected before the item is put into service and 10 showing a very low probability of detection. In Table 7.1 the 'bracket fails' failure mode of the tie bar bracket is rated at $1 \times 8 \times 5 = 40$ whereas the 'HD bolts loosen' failure is rated at $5 \times 5 \times 8 = 200$, indicating that tightening the fixing bolts is more of a potential problem.[1]

Reliability

The techniques used for safety and reliability analysis are not solely concerned with the prevention of accidents. They also have an important economic value, for they can be utilised to evaluate the financial consequences of equipment failure and disrupted production. Analysis of the probability and the costs of corrective modifications help management to make cost-effective decisions about preventive maintenance repairs and replacement of plant.

Safety cases

Risk analysis, fault trees and FMEA are also used in the preparation of safety cases. The Flixborough explosion of 1974 (Fig. 7.3) focused atten-

Fig. 7.3. Flixborough disaster, 1974

tion in the UK on the problems of major hazards. Other major industrial accidents in Europe stimulated the European Community to prepare in 1977 a directive on major accident hazards. This directive is implemented in the UK by the Control of Industrial Major Accident Hazards (CIMAH) Regulations 1984.

A manufacturer who has control of an industrial activity defined in the CIMAH regulations is required, among other things to submit to the HSE a written report known as a safety case, which demonstrates that the hazards and risks from the industrial activity have been identified and that the appropriate safeguards have been analysed and found acceptable. The safety case

- identifies the nature and scale of use of dangerous substances
- places the site in its geographical and social context
- identifies the type, relative likelihood, and consequences of potential major accidents
- identifies the control regimes and systems for the site
- identifies whether the manufacturer or occupier of the site has considered the adequacy of the controls.

Whereas much of the preparation of safety cases is carried out by mechanical, electrical and chemical engineers, it is desirable that civil engineers understand the principles, partly because of the civil and structural engineering elements of industrial and nuclear sites, and partly because of the relevance to planning and site investigation for new projects that fall within the regulations.

8

Information

The need for information on safety

The requirement of section 2(2)(c) of the HSW Act that 'employers are to provide such information as is necessary to ensure the health and safety at work of his employees' is a very onerous one in construction. The industry is so diverse that almost every type of accident and environmental health hazard is possible for its employees. To ensure that information reaches those who face such a multitude of hazards requires good organisation and dedication.

In addition, it is fairly evident that information is necessary to advise engineers, technicians, architects and all those who are in a position to adversely affect the health of construction workers and members of the public by their acts or omissions. Many of the obvious hazards are not covered in university degree coursework or in textbooks and means have to be found to provide professionals with the knowledge they need to avoid repetition of problems experienced by others.

Information is the means of communicating the art of good safety practice to all employees. It may be either advisory or mandatory. Advisory information is available in publications, posters and films. However, information sheets or practice notes and mandatory procedures will invariably have to be prepared in-house to give proper emphasis to the particular company's procedures and requirements.

Information prepared in-house

The primary medium for making safety information available is print. It takes the form of handbooks, standing instructions, practice notes and advisory notes prepared for or by the company safety adviser for distribution to those for whom it is intended.

Organisation for preparation and circulation

In companies with more than a few employees there are several subjects — health and safety is only one — that require the circulation of information to employees. It is common practice for the person responsible for administration to establish a routine for achieving this.

A system that operates in many companies is that the authors of instructions or practice notes send the completed and approved texts to the administration department for printing and circulation. While many small firms rely on less formal systems for circulating essential information, the principle that information about health and safety matters should reach those for whom it is intended by design, rather than by chance, remains the same.

Safety handbooks

To ensure that all staff and operatives whether employed by contractors, client organisations or professional firms are aware of the hazardous nature of the industry, it is recommended that a company safety handbook is produced and given to employees when they are first appointed. Some companies include chapters on safety and welfare within a general staff handbook but a separate publication is preferable. Since there are some procedural matters that are common both to safety and general management, for example accident reporting, it is important for the staff handbook and the safety handbook to be compatible.

The purpose of a company safety handbook is three-fold. First, it sets down briefly and simply the responsibilities of the employer under the HSW Act. Second, it tells the employees what their duties are and, third, it outlines the hazards of the job and how they should be avoided. Because of the diverse nature of some larger companies the third function, that of describing the hazards, is contained in a separate booklet or in guidance leaflets.

A typical safety handbook

Safety handbooks may contain a synopsis of the company's health and safety policy statement and a personal message from the chairman (*see* pp. 143 and 145). The following is an example of a typical safety handbook for a firm of civil engineers where the policy statement is published elsewhere.

General Introduction

The Health and Safety at Work etc. Act 1974 requires employees to safeguard their own health and safety and to co-operate with their employers in ensuring that the wellbeing of other company employees and the public at large is not put at risk.

This booklet outlines the duties under the Act that govern the responsibilities of the employer and employee and provides some pointers to safety in circumstances commonly encountered in our business. Please read it carefully and keep it in a safe place for reference purposes.

While the Act is directed only to Great Britain (another similar Act applies to Northern Ireland), its principles should be taken to apply wherever company personnel are operating.

Employer's Responsibilities

The duties of employers in complying with the Health and Safety at Work etc. Act 1974 are outlined in the company health and safety policy statement that is attached to principal notice boards. The statement also describes the company policy on health and safety together with the organisation set up to implement it.

The Act places demands on both employer and employee in order to achieve its principal aims of

- securing the health and safety of people at work, and
- protecting others from hazards arising from work activities.

The Company, for its part will

- provide you with a safe working place commensurate with the nature of your job
- inform you of any hazards connected with your job and train you to safeguard against them
- provide you with protective clothing and safety equipment if your job requires it
- maintain procedures and systems of work to assist you in carrying out your job with minimum risk of damaging the health of others
- support any reasonable contentions you may make that the conditions of any location in which you are required to work are hazardous beyond the nature of your job.

Employee's Responsibilities
For your part you are required to

- take notice of health and safety notices and literature
- know your fire drill: obey fire instructions and office evacuation procedures
- know what to do in the event of an accident and, where called on, do it
- wear clothing and footwear suitable for your job
- wear protective clothing and use safety equipment where your task is hazardous or if you have been instructed to do so.
- ensure that, by your acts or omissions, you do not put at risk the health and safety of your fellow employees, other people at work or anyone else
- avoid misusing anything provided for your safety
- attend the training courses provided for you
- obey the health and safety instructions in force at any site or other establishments you may visit in the course of your employment with the company
- be mentally and physically fit for the work you have to do.

General safety information for employees

It is usual for the company safety handbook to contain examples of the principal forms of accident in the construction industry, some of the environmental hazards relevant to employees of the company and the precautions to be taken. Another section usually contains details of the protective clothing and safety equipment available to employees, information and training, and first aid and reporting procedures in the event of an accident.

Some large organisations such as British Rail require compliance with safety rules particular to their type of work. These need to be included within the safety handbook or, better still, as a separate booklet that may be issued to all contractors and consultants who work for them as well as to their own employees. (Note the HMS *Glasgow* case, pp 20 and 21.)

Company safety literature for managers

Managerial instructions vary from company to company. They may be referred to as standing instructions, practice notes, procedural notes or management memoranda. There is a requirement for several grades of staff to receive company instructions on their duties in regard to health, safety and welfare. Managers are not only employees but they also represent the employer to all staff and operatives who report to them. It is therefore necessary to ensure that they are thoroughly aware of the duties imposed on them by law and company policy. Staff who require instructions on health and safety include divisional and departmental managers employed by authorities, boards and major companies, construction managers and site managers employed by contractors and project managers and resident engineers employed by consulting engineers. The subject matter of the managerial instructions would include the terms of reference and duties described in chapter 5.

Practice notes. Most authorities and companies involved in construction activities find it necessary to prepare and circulate information sheets on procedural or technical matters. Examples of procedural matters that might be covered on information sheets include arrangements for safety committees and safety representatives; arrangements for protective clothing and equipment; accident reporting procedures; and permit-to-work systems.

Information sheets and guidance notes on most construction hazards are already available in printed form from HSE and other sources. However, they are often too general for distribution to all employees, and companies need to explain in-house procedures, giving necessary information to ensure that decisions are made at the right managerial level. Thus it is necessary for companies to prepare practice notes describing hazards that particular employees have to face and the precautions to be taken. A few examples of such practice notes that have been prepared by various contractors for their own use include those for

- working in sewers and confined spaces
- safety precautions during prestressing operations
- safety precautions when working on live highways
- working near live underground and overhead services
- precautions when surveying old and unoccupied buildings.

An organisation must be set up within the company not only to distribute the practice notes but also to ensure the regular updating both of the contents and the circulation lists. Examples of instructions, practice notes, a safety handbook and site notice board poster prepared in-house are shown in Fig. 8.1.

Posters

A number of subjects are more suited for display on notice boards than for circulation to individual staff members. These include local arrangements for health and safety, including the names of safety representatives, first-aid officers and the procedures for reporting accidents.

Fire and evacuation procedures and particular hazard warnings are other subjects suitable for posters which may be either specially produced or obtained from the Royal Society for the Protection of Accidents and the

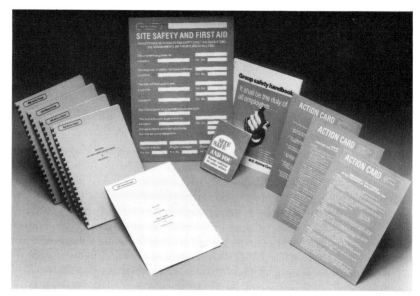

Fig. 8.1. Examples of instructions and guidance notes

British Safety Council. The range of the posters is continually extended, some subjects are updated or deleted and catalogues of currently available posters may be obtained from the organisations concerned. Examples of posters are given in Fig. 8.2.

Publications

Some publications specialising in safety and health are suitable for wide circulation throughout the staff membership, while others are suitable for reference, for lending libraries or for retention by managers and safety advisers. The specialism and size of the company will determine the way in which safety information is made available to staff.

Publications produced by the HSE

A comprehensive range of free and priced publications is produced by the Health and Safety Executive. A list of HSE publications and films relevant to the construction industry may be obtained from the HSE Prestel programme or from the Health and Safety Executive Library and Information Service in Sheffield, Bootle or London (*see* Appendix 3). A list of HSC/E publications relevant to construction is given in Appendix 4.

Other guidance notes

Guidance notes and codes of practice for the use of contractors and other who may be putting themselves and others at risk include

- *Recommendations on the avoidance of danger from undergound electricity cables*, National Joint Utilities Group

Fig. 8.2. Examples of safety posters available from RoSPA

- *Gases in tunnelling — guidance notes on strategies and safety*, North West Water
- *Code of practice for safe working in the vicinity of British gas transmission pipelines*, British Gas
- *Track safety handbook*, British Rail
- *Notes for guidance in relation to personnel who work on motorways and trunk roads*, Department of Transport and the Country Surveyors Society (available from Department of Transport, London).

Construction safety manual

A manual published by the Building Advisory Service of the Building Employers Confederation in Birmingham, *Construction safety*, is regarded as the standard reference work and training manual for those in the construction industry. It contains over 30 chapters ranging from legal and organisational matters to specific safety topics and is published in two volumes in loose-leaf format so that updated pages can easily be substituted. It is necessary for employers to ensure that their copies of *Construction safety* are kept up to date and those with several copies are advised to give them catalogue numbers, and arrange for their whereabouts to be recorded by a librarian or administrator, who will be able to call the sets in for revision when new or revised pages are issued by the publishers.

Other publications available from the Building Employers Confederation (BEC) on safety and health include

- *Guide to the Construction Regulations*, BEC and Federation of Civil Engineering Contractors (FCEC)
- *Supervisors' safety booklet*, FCEC
- *Site safe and you*, National Joint Council of the Building Industries
- *Site safety supervisors compendium*, BEC and Building Advisory Services
- *Scaffolders and users guide to safe access scaffolding*, National Federation of Scaffolding Contractors (NFSC)
- *Crane riggers and users guide to safe suspended platforms*, NFSC
- *Guidance note on safety responsibilities for subcontractors on site*, BEC

Safety manual for mechanical plant construction

A safety manual for mechanical plant construction is published by the Oil and Chemical Plant Constructors Association, and is a loose-leaf publication in two volumes. It is suitable for engineers involved in the petrochemical industry.

Books on health, safety and welfare

Books on health, safety and welfare useful to those in the construction industry include

- *RoSPA health and safety practice*, by J. Stranks and M. Dewis, Pitman
- *Construction hazard and safety manual*, by King and Hudson, Butterworth
- *Health and safety in factories (Redgraves)*, by Fife and Machin, Butterworth
- *Croner's guide to health and safety*, Croner
- *Health and safety at work*, by M. Goodman, Sweet and Maxwell
- *Construction Regulations handbook*, RoSPA

Magazines and journals

Staff and particularly middle managers responsible for safety should be encouraged to read one or more of the journals produced by safety organisations. Members of the Institute of Safety and Health receive *The safety practitioner*, which is also available to non-members. Members of RoSPA receive the monthly RoSPA bulletin and *Occupational safety and health*, also available to non-members. Subscribers to the British Safety Council receive *Safety management* monthly. *Health and safety at work* is a widely read monthly magazine published by EMAP Maclaren Ltd. Circulation of the journals to a wide readership among management and staff is a good cost-effective way of improving their safety awareness.

For safety advisers and others with managerial responsibility the *Construction service information* bulletin is available from HSE. Also, *Construction safety information service* is available by subscription from the Building Advisory Service of the BEC which supplies abstracts of articles and other matters relating to construction safety. Those interested in safety and reliability may wish to consult the *Safety and Reliability Society Journal*. *Site safe news*, published by the HSE twice a year and suitable for wide circulation on construction sites, is available from the Construction National Industry Group.

Films

Films and videos relating to the construction industry are available for hire or purchase in 16mm or in VHS or BETA format. Those that are specially useful for training purposes include

- *No questions asked*
- *Eyes down*
- *Acts and omissions*
- *Watch that space: confined spaces in the construction industry*
- *Safety in construction: scaffolding*
- *Checkmate I*
- *Checkmate II*
- *A site safer*
- *Hangman*
- *All in a day's work*
- *The Contract*

Details may be obtained from: Rank Millbank Training, 3 Centaurs Business Park, Syon Lane, Isleworth, Middlesex TW7 5QD; Live Action Communications, 113 Humber Road, London SE3 7LW; RoSPA Film Library, Cannon House, The Priory, Queensway, Birmingham B4 6BS; CFL Vision, P.O. Box 35, Weatherby, West Yorkshire LS23 7EX.

Many large organisations make in-house films and videos for staff safety training and for contractors who work on their establishments. Some films and videos on the more popular safety topics such as action in the event of a fire are available from public libraries and also some excellent fire films are available on free loan for training purposes. Enquiries may be made to the Fire Protection Association in London, or to any county fire brigade.

9

Training

The need for safety training

The general duty of the employer to 'provide such instruction and training as is necessary to ensure the health and safety at work of his employees' and, presumably, others who may be affected by their acts or omissions, leaves little doubt that one of the first questions to be asked at an enquiry after an accident is, 'Have the staff and operatives involved been properly trained?' To satisfy this general duty, employers in the construction industry find it necessary to provide training at several levels from a basic introduction to health and safety for everyone in their employment to safety training courses for managerial staff. Moreover, safety should form part of all training of construction operatives.

Induction courses

Many companies in the construction industry provide induction courses for all new employees. The courses include an outline of the organisation and management of the company, personnel and welfare arrangements and staff facilities available. The opportunity must not be missed to include a talk on health and safety. Points that must be stressed include

- they must read their company safety handbook
- heeding their personal safety and the safety of their colleagues and the public at large is a requirement not only of the law but of their employment with the company; breach of that duty can result in a reprimand or, in extreme cases, dismissal
- how to obtain first aid
- what to do in the event of a fire
- what to do in the event of an accident
- how to evacuate the building
- how to obtain protective clothing and safety equipment
- a brief outline of the likely hazards employees may encounter, together with examples of accidents that have occurred
- a brief outline on further safety training and information that is required for managers and personnel on site.

It is recommended that the basic introduction to health and safety should last for at least two hours and should include visual aids, slides and possibly a film.

Graduates and technicians

It is recommended that time is devoted at university or college of technology to the subject of health and safety in the construction industry. The training should cover an outline of safety legislation and the duties of employers and employees. It should describe the hazards that professional staff face in the course of their work and outline the problems facing designers in ensuring that their work is safe to build, operate and maintain. Emphasis should be given to the practical ways in which graduates can help to improve the poor safety record of the construction industry.

The subject must be expanded on during the postgraduate training period. All civil engineering graduates and other professionals should have formal lectures in health and safety legislation, site safety and safe systems of working. At their professional examinations graduates and technicians are expected to satisfy the examiners regarding their knowledge and experience in basic health, safety and welfare practice.

Managers

It is insufficient for senior staff merely to have attended the basic induction course and graduate training in health and safety. Their duty as representatives of their employer has to be made quite clear to them and it is therefore desirable that all managers attend a two- or three-day course on management in construction. The syllabus for the course depends, to some extent, on the branch of the industry to which the particular employer belongs.

Items on such a course should cover

- the law: health and safety legislation applicable to the construction industry, and the law of contract and tort with special reference to duties and responsibilities for safety
- safety policies: the policy of the particular employer and methods of implementation, policies of other authorities, contractors and consultants, and the role of the company safety adviser
- hazards: the principal causes of accidents on site, and occupational health hazards
- accident prevention: application of the Construction Regulations and other relevant regulations and codes
- accidents: procedures in the event of an accident, and accident reporting.

Specific safety training items will need to be added to the syllabus depending on the branch of engineering of the middle managers attending the course. For example, construction and site managers need to cover in greater depth the regulations, notices, records and test certificates that must be complied with. In particular, they need to have a thorough understanding of the Construction Regulations and other regulations and codes relevant to their particular line of business, and the ways in which the requirements can be met.

Project directors, partners and managers with professional firms need to understand how the company meets its statutory duty to provide for all the safety of its employees especially when away from the office. In addition, they need to cover

- statutory and common law duties relating to designers and specifiers
- company quality procedures and their implementation
- contractual relationships with clients and contractors on safety matters
- company procedures for providing safety information and training for and supervision of its employees.

Managers employed by government, authorities and other major employers need to cover the safety aspects of their own industry and how their particular organisation enforces its safety procedures in relation to its own employees and outside contractors.

Safety advisers

Where a safety adviser or safety officer is employed full time it is recommended that he attend a course leading to qualification as a member of the Institution of Occupational Safety and Health. The training programme recommended for construction safety advisers is specially developed by the Building Employers Confederation, the Federation of Civil Engineering Contractors, the Construction Industry Training Board (CITB) and the Institution of Occupational Safety and Health, with the help and co-operation of the Health and Safety Executive. Full details of the scheme of training are given in CITB Training Recommendation 13, and details of the courses may be obtained from Brooklands School of Management in Weybridge, Surrey.

Resident engineers

The particular situation of resident engineers, or Engineer's Representatives under the ICE Conditions of Contract, should be covered by a training course of at least two days with a syllabus drawn from elements of the middle management training syllabus. The resident engineer has to be thoroughly aware of his responsibilities under the HSW Act and how to relate this to his duties under the Contract. The engineer's safety adviser, appointed to assist the resident engineer in day-to-day safety matters, should attend the same training course. Details of courses specifically arranged for resident engineers and project managers with site supervisory responsibilities are obtainable from the Building Advisory Service of the BEC in London or from W.S. Atkins in Epsom, Surrey.

Other training courses

Courses in safety for site management are held regularly at the CITB National Training Centres at Bircham Newton and Glasgow and on demand at Birmingham. The courses, which are recognised by the Institution of Civil Engineers for continuing education and by the Chartered Institute of Builders, are of one week duration, and on successful completion, the site manager's safety certificate is awarded.

The many courses run by the CITB at their four national training centres at Bircham Newton, Glasgow, Erith and Birmingham include elements of safety. In addition to those mentioned above there are courses for safety supervisors on specialised subjects such as timbering, cranes and scaffolding and also on the subject of safety training. Details of these courses may be obtained from CITB, Bircham Newton, or from the other national training centres or regional offices.

For safety advisers or managers wishing to gain a basic knowledge of safety in the construction industry a course in safety for middle management may be suitable. Such a course run by the RoSPA Occupation Safety Training Centre in Birmingham is designed to acquaint delegates with their responsibilities under the HSW Act and with related legislation, and to inform them of safe working practices. For details of these and other courses at the centre apply to the Training Centre Manager, RoSPA, Birmingham.

In addition to those mentioned previously a number of safety training centres in Britain and Northern Ireland cover the needs of companies who do not have sufficient resources to run in-house courses. Details of these may be obtained from the Federation of Civil Engineering Contractors, the Building Employers Confederation or the Health and Safety Executive.

For first-aid training in accordance with the Health and Safety (First Aid) Regulations 1981, courses held by the St John Ambulance, St Andrews Ambulance Association and the British Red Cross Society (*see also* p. 224) are approved by the Health and Safety Executive.

Training in fire prevention and action in the event of a fire may be obtained by application to any county fire brigade or to the Fire Protection Association in London. The fire brigade also give training in confined space entry and rescue, including the wearing and maintenance of breathing apparatus.

10

Protective clothing and safety equipment

General requirements

The wearing of protective clothing and the use of safety equipment is crucial in the battle to reduce the effects of accidents on construction sites. Employers have a legal duty to provide protective clothing and equipment free of charge for certain operations. Even where there is no statutory requirement to provide protective clothing, employers have a duty to ensure that conditions are safe and without risk to health and this duty may often be discharged by the provision of safety clothing for their employees. Employees also have a duty to preserve their own health and safety and it is in everyone's interest that they should be adequately clothed whether special clothing is provided by the employer or not.

Disciplinary action may be taken against employees who refuse to wear the protective clothing supplied by their employers. The refusal to wear protective clothing was decided by the court to be a dismissable offence in a specific case: *Marsh* v. *Judge International Housewares Ltd* [1976].

However, where letters of appointment clearly state that the wearing of protective clothing and use of safety equipment as instructed is a condition of employment and that any refusal to do so is subject to reprimand or in extreme cases dismissal, employees are left in no doubt about their obligations. Employees should also be informed that they have a legal duty not intentionally to interfere with or to misuse anything provided for their safety.

Mandatory protective clothing and equipment

Employers are required by law to provide

- adequate and suitable protective clothing for all people required to work in the open air during rain, snow, sleet or hail (Construction (Health and Welfare) Regulations 1966)
- approved protective breathing equipment and protective clothing for people at work where asbestos dust may be present in the atmosphere where it cannot be removed by exhaust ventilation (Asbestos Regulations 1969)
- insulating boots and gloves for electricians (The Electricity Regulations 1944)
- goggles, visors, spectacles and face screens to protect the eyes where

specified processes are being carried out (Protection of Eyes Regulations 1974)
- safety belts, harness, lines, etc. where it is not practicable to provide safe working platforms or safety nets (Construction (Working Places) Regulations 1966)
- ear protectors where it is not practicable to reduce noise exposure to below prescribed levels (Noise at Work Regulations, 1989)
- safety helmets (Construction (Head Protection) Regulations, 1989).

Standards of personal protective equipment

It should not be assumed that all protective safety wear is satisfactory for all circumstances. In general, the selection of protective clothing should take into account

- the ability of its material to resist penetration
- the adequacy of its design
- the environment in which it will be worn
- for protection against dust, its dust-release characteristics.

Leaflet IAC L16 prepared by the Construction Industry Advisory Committee (CONIAC) and published by the Health and Safety Commission (*see* Appendix 4) gives advice on protective clothing and footwear in the construction industry. It summarises the general principles to be considered by management for the selection of protective clothing and lists the protection afforded by the materials from which the clothing may be manufactured.

Those responsible for providing protective clothing should

- identify the risks before work starts
- remove risks at source where possible
- where hazards cannot otherwise be controlled, consider protection against physical injury, chemical injury, ill-health, irritancy or nuisance, electrical risks, temperature and humidity extremes and wet conditions
- make sure that protective clothing complies with relevant standards
- consider the compatibility of clothing and other equipment (e.g. respirators, life jackets, helmets)
- consult employees
- provide information and training in its use
- provide for cleaning, maintenance and storage.

The wearing of protective clothing by professional staff

Whatever the legal requirements, it is essential that professional staff and other managers in the construction industry always wear safety helmets, protective footwear and other clothing, both for their own protection and as an example to the workforce. Visitors and those on short-term or *ad hoc* assignments are prone to ignore safety protection, often because of lack of managerial control, but they are equally vulnerable and should not enter hazardous workplaces unless they are properly protected.

Protective clothing for use on construction sites

The range of available protective clothing and equipment is wide and there are many manufacturers and retailers. We are concerned mainly with the protective clothing and safety equipment used in the construction industry and in particular with that commonly worn and used by civil engineers and other professionals in the course of their site duties.

Several contractors and major authorities have arrangements with retailers which enable employees to select clothing from a number of styles and sizes. Where the clothing is not supplied by the company free of charge, purchase at a discount is often available. Some suppliers have mobile showrooms that visit customers on a regular basis.

Safety helmets

Legislation to enforce the wearing of safety helmets on construction sites is embodied in the Construction (Head Protection) Regulations 1989, which came into force on 30 March 1990. Safety helmets should be supplied and worn unless circumstances are specifically exempt from the regulations.

Safety helmets must comply with BS 5240: 1987 Industrial Safety Helmets, and have an adjustable plastic or terylene head harness which can be removed for cleaning. A typical recommended heavy-duty helmet is shown in Fig. 10.1. It will take slot-fix accessories such as chin straps, ear defenders and face shields.

Fig. 10.1. Heavy-duty helmet (courtesy ARCO Ltd)

PROTECTIVE CLOTHING AND SAFETY EQUIPMENT 211

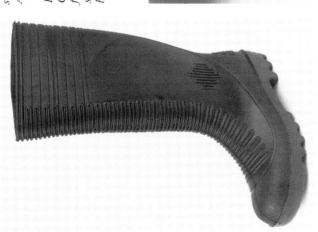

Fig. 10.2 (left). Construction industry safety wellington boot (courtesy ARCO Ltd)

Fig. 10.3 (below and right). Construction industry safety boots: (a) flexible chukka boot; (b) heavy duty site boot; (c) heavy duty commando boot (courtesy ARCO Ltd)

Safety footwear

All workers, including those in offices, should wear sensible shoes or boots to prevent abrasions and cuts from sharp objects on the ground. Construction sites are especially hazardous. Nails, reinforcing bars, binding wire and other sharp objects on rough uneven and untidy ground result in thousands of foot accidents, some of which become very serious. Recommended wellington safety boots for wet or muddy conditions have a steel midsole, to resist penetration from upstanding sharp objects, as well as a steel toecap. They are shown in Fig. 10.2 and are distinguished from the ordinary heavy-duty wellington boot by a red coloured sole.

The heavy-duty site boots with steel midsoles illustrated in Fig. 10.3 are recommended for general site wear. Combat boots and chukka boots are somewhat lighter (Fig. 10.3(a)) and are suitable for use where the risk of sole penetration is less but not on constuction sites.

Hearing protection

The noise levels in some areas on construction sites are often well above the level which will cause permanent hearing damage to workers in the vicinity. Many people are in fact unaware that their hearing has already been damaged as a result of work or leisure activities and are therefore less susceptible to loud noise and more resistant to wearing protection (*see also* p. 136).[1]

The common form of hearing protection in the industry is the ear defender consisting of a headband and cup. There are several types of headband depending on helmet attachment and hairstyle. More important, however, is the choice of cup, which must be selected for the appropriate noise level recommended by the manufacturer. Choice of an incorrect protection will not reduce the sound level by the required amount. Helmet muffs are also available (Fig. 10.4). These can be snapped

Fig. 10.4. Helmet muffs for hearing protection (courtesy ARCO Ltd)

Fig. 10.5. Eye protectors (courtesy ARCO Ltd)

out into the rest position or swung up on to the helmet when not in use. Ear plugs are also widely used in the construction industry.

Eye protection

The dangers from flying particles and dust are obvious to most construction workers and the wearing of safety eyewear more readily accepted than other safety protection. A number of processes common in the construction industry that cause flying fragments or particles, chemical splash, intensive glare, etc. are those for which eye protection is required under the Protection of Eyes Regulations 1974. The regulations apply to those whose eyes may be injured, e.g. supervisors and inspectors, as well as the operatives.

Goggles, face shields and spectacles are available giving protection against impact, dust, chemical, molten metal, and gas hazards, and managers should ensure that the correct protection and its correct grade is used by the employee (*see* Fig. 10.5). Eye protectors conforming to BS 2092: 1987 *Specification for eye protectors for industrial and non-industrial use*[2] are marked to indicate the protection they are certified to give. Inspectors of welding need to wear welding goggles with lenses approved to BS 1542 and BS 679.[3,4]

Respiratory protection

The choice of protective equipment against the effects of dust, mist, fumes and lack of oxygen is wide, and great care should be exercised in its selection if the health of the wearer is not to be seriously damaged. It is convenient to consider, firstly, protection against contaminants in the atmosphere, and secondly, protection against oxygen deficiency.

Respirators. A number of respirators are available for protection against dusts, vapours and fumes. They have face masks to cover the nose and mouth or nose, mouth and eyes and are fitted with renewable filters according to the protection required. Powered respirators, usually with a full face mask, have filtered air supplied at positive pressure by motor, powered by rechargeable batteries.

Disposable maintenance-free respirators or face masks that meet the requirements of BS 6016: 1980 *Specification for filtering facepiece dust protectors*[5] type 1, type 2 and type 3 are suitable for fine dusts; dust and mists; and dust, mist and fumes, respectively. Many other types of disposable respirator are made, for protection against the effects of spray paint, organic vapour, acid gas, ammonia, etc. and nuisance odours.

All respirators should indicate the protection that they provide, to prevent incorrect use. Even though they may comply with the relevant British Standards, they must also be approved by the Health and Safety Executive.

Breathing apparatus. Those entering confined spaces to which it is not practicable to provide an ample supply of fresh air, or other atmospheres deficient in oxygen or dust-laden environments which cannot be dealt with by ventilation, must wear breathing apparatus. Standard, positive pressure, compressed-air breathing apparatus is shown in Fig. 10.6. Staff must

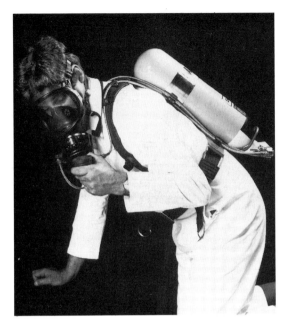

Fig. 10.6. Compressed-air breathing apparatus (courtesy ARCO Ltd)

be properly trained in its use and most fire brigades are prepared to give this training. One other person must always be present as a precaution in case of difficulty. The compressed-air cylinders have varied working durations (e.g. 20–35 min), depending on their capacity, and arrangements have to be made to recharge them.

A system that gives the operative more manoeuvrability and somewhat greater protection is the airline respirator, which is fed by airline from cylinders mounted on a trolley unit (*see* Fig. 10.7). This incorporates a whistle warning unit and up to 90 m of hose.

Outer wear

For general site wear by professional staff in the UK the parka jacket is popular. It is warm and waterproof and strong. Donkey jackets are also popular for supervisors and tradesmen. Both are shown in Fig. 10.8. For those required to work in cold and wet conditions, a range of inclement weather clothing, either two-piece or overalls, is available from many manufacturers.

High-visibility garments

A variety of high-visibility jackets, tabards and belts is available for use on public highways, railways, airports, industrial premises and construction sites where workmen and staff need to be seen by drivers of vehicles and plant. Garments for use on the highway should conform to BS 6629: 1985 *Optical performance of high-visibility garments and accessories for use on the highway.*[6] Examples are shown in Fig. 10.9.

PROTECTIVE CLOTHING AND SAFETY EQUIPMENT   215

Fig. 10.7. Airline respirator (courtesy ARCO Ltd)

(a)

(b)

Fig. 10.8. (a) Parka jacket; (b) donkey jacket for site wear (courtesy ARCO Ltd)

Fig. 10.9. High-visibility garments (courtesy ARCO Ltd)

Other protective clothing

Many items of clothing and equipment worn by construction workers and staff are not described in this chapter. Further information may be obtained from a catalogue of a specialist supplier.

Other safety equipment

Engineers often have to work at heights or go underground. The hazards of heights and confined spaces can never be entirely eliminated by the normal provision of safe working places so personal protection must be worn. Specialist safety equipment for mountaineering, pot holing and caving is used by some engineers for inspecting or surveying high structures or sewers and tunnels and, of course, diving equipment is used for underwater work. However, only the most common equipment is described here.

Safety belts and harnesses

People required to work at a height where the provision of safe working platforms is impracticable should wear a safety belt or harness and line or lanyard attached to a reliable strongpoint above the working position. Full safety harness is preferable to a safety belt because injury is less likely in the event of a fall.

Harnesses should be fitted with a lanyard, which limits any free fall to 600 mm, or up to 1.2 m provided the lanyard embodies an integral shock absorber. The lanyard for use with a chest or general-purpose harness should allow for a maximum 2 m fall from the point of attachment. Types of safety belt and harnesses are described in BS 1397 Specification for industrial safety belts, harnesses and safety lanyards. The use of safety belts and harnesses in the erection of structural frameworks is described on page 79.

Chest or body harnesses, not safety belts, should be worn by those entering confined spaces because, when rescue is necessary, an unconscious person needs to be in an upright position to pass through the exit manhole. Lifting an unconscious person wearing a safety belt causes the victim to double over, and rescue through a manhole is then extremely difficult.

11

Accidents and first aid

The accident rate in the construction industry is such that most professional engineers and builders will be involved with an accident at some time in their career. It is therefore in the interests of all engineers, the unfortunate victims and their relatives and friends that professional staff should be fully prepared to deal with accidents when they occur, to provide first aid and to carry out the proper investigation and reporting procedures afterwards.

Site arrangements for health, safety and welfare
Site manager's duties

When work starts on a new construction site, the contractor has to carry out a number of duties that are required by law. In practice, they are usually performed by the site manager.

One of the site manager's first duties is to notify the local office of the Health and Safety Executive within seven days of the start of building operations or works of engineering construction on Form F10 unless the work is expected to last less than six weeks. He may also request a visit from the Factory Inspector to discuss site safety provisions and problems. The HSE should be informed on Form OSR1 before occupation of any offices that are to be in use for more than six weeks where fixed (or six months where mobile) where the offices are to be occupied for more than 21 man-hours per week.

It will also be necessary to make an application to the HSE for a fire certificate in cases where offices on construction sites are occupied by more than 20 people on the ground floor or ten people above the ground floor, or where explosives or flammable materials are stored in or below the premises.

The site manager is recommended to inform the local fire brigade and ambulance service of the location of the new site and its points of access so that in the event of a 999 call for fire or rescue services or for an ambulance the services will be familiar with the site. He may also request a meeting on site to discuss safety and first-aid provisions.

The site manager must make sure that posters are displayed in prominent places on the site and at site offices, indicating the action to be taken in the event of an injury or dangerous occurrence. A typical site notice is shown in Fig. 11.1.

ACCIDENTS AND FIRST AID

Telephones

For reasons of safety at the very least, a site telephone should be installed at the earliest opportunity. The first few days on site are no less hazardous than any other time. Until a telephone line is installed it is recommended that a portable telephone be provided. The emergency services telephone number (999), the emergency numbers of key site personnel and the number of the local office of the HSE should be prominently displayed near the telephones. Portable telephones should always be provided where no fixed telephone line is practicable, e.g. on site investigation contracts, utilities contracts, minor roadworks and surveys.

Emergency procedures plan

When the site work includes operations that might be more hazardous than normal a plan should be drawn up by the contractor for dealing with

```
                    Constructwell Limited
                    [                    ]
              Site arrangements for health, safety and welfare

    Site manager            [          ]         Tel. [          ]
    Deputy site manager     [          ]
    Site safety supervisor  [          ]         Tel. [          ]
    Safety representatives  [          ]
                            [          ]
    First aiders            [          ]
                            [          ]
                            [          ]

              The first-aid room is located at [                    ]
    The nearest hospital is [                                        ]

              In case of an accident or fire phone [          ]
                    or dial [ 999 ]   at public phone

    The telephone number of the local office of the
    Health and Safety Executive is                  [          ]

    Constructwell Limited:
    The director responsible for this site is        [          ]
    The construction manager responsible for this site is [          ]
    The Company safety adviser is                    [          ]
```

Fig. 11.1. A typical site notice

any foreseeable major incident. The plan should include methods of evacuation and rescue of all people likely to be on site and should be formulated following discussion with the emergency services (police, fire and ambulance) and other relevant site organisations, e.g. subcontractors, resident engineer and client organisation. The plan is called an emergency procedures plan. Evacuation or rescue or both are particularly difficult where the incident affects men working

- at height
- in confined spaces
- in sewers, particularly live sewers
- in tunnels being constructed where the length of blind heading is over 500 m
- over or adjacent to water
- in remote areas with difficult access.

Preparation of an emergency procedures plan leads to consideration of a number of items, depending on the severity of any incident that is likely to occur

- provision of training for the contractor's rescue team
- discussions with police, fire brigade and ambulance service on such matters as
    - training by the fire brigade in the use of breathing apparatus
    - traffic control in the event of an accident — access routes and assembly areas
    - clarification of any special rescue equipment that may be required and who should provide it
    - familiarisation visits by the emergency services — rescue exercises
- the use of a tally procedure or clock on–clock off system to account for everyone on the site
- involvement of site medical staff or first aiders in any major incident
- how an alarm will be given
- approval of the plan by all concerned, including the emergency services, followed by its circulation and publication in prominent positions.

The emergency procedures plan may be accompanied by an event tree, an example of which is indicated in Fig. 11.2.

First-aid facilities
Legal requirements

The Health and Safety (First Aid) Regulations 1986 require employers to provide or ensure the provision of appropriate equipment and facilities enabling first aid to be rendered to their employees if they are injured or become ill at work. Employers must inform employees of the arrangements that have been made for rendering first aid, including the location of equipment, facilities and personnel. Reference should be made to the HSE approved code of practice *First aid at work*.[1]

ACCIDENTS AND FIRST AID   221

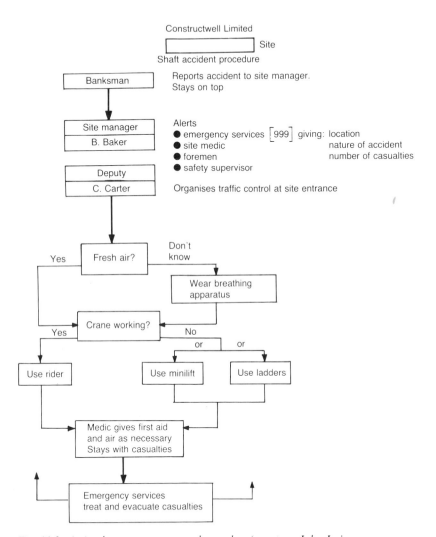

Fig. 11.2. A simple emergency procedures plan (courtesy John Laing Construction Ltd)

The guidance note on first aid at work (HS(R)11) prepared by the Health and Safety Executive gives advice on equipment and facilities, first aiders and training, and contains specific information for small establishments. A useful leaflet, *First-aid provision in small workplaces — your questions answered* (IND(G)3(P)) is published by the Health and Safety Executive.

Provision for self-employed people

Under the regulations, a self-employed person must provide equipment, or ensure that equipment is provided, to enable him to render first aid to himself at work.

First-aid rooms

On all large construction sites and others where access to accident and emergency services is difficult, or where there is dispersed working, a suitably equipped and staffed first-aid room should be provided (Fig. 11.3).

- should be in the charge of a suitably qualified person
- should be readily available and used only for first aid
- should be of sufficient size and located at a point of access for transport to hospital
- should contain suitable facilities and equipment, be ventilated, heated, lighted, cleaned and maintained, and a waiting room should be provided
- should contain a sink with hot and cold running water, soap, nailbrush and paper towels, drinking water and suitable vessels, smooth-topped impermeable working surfaces, a couch with pillow and blankets, clean garments for use by the first aider, an adequate supply of sterile dressings, a clinical thermometer, a store for first-aid materials and a lidded refuse container
- should be equipped with telephone, radio and siren for immediate contact with the first aider on call, and the ambulance and rescue services.

First-aid boxes

Regardless of the number of employees there must be at least one first-aid box on any construction site or work location. The first aiders must have easy access to first-aid equipment and each box should be placed in an easily accessible location. The boxes should contain nothing but first-aid equipment and be cleaned and checked regularly. The contents of first-aid boxes depend on the number of employees and are listed in Appendix 5.

Where employees work singly or in small groups small travelling first-aid kits should be provided and employees should be instructed in basic first aid. The contents of the kit should include at least

- six sterile dressings

ACCIDENTS AND FIRST AID

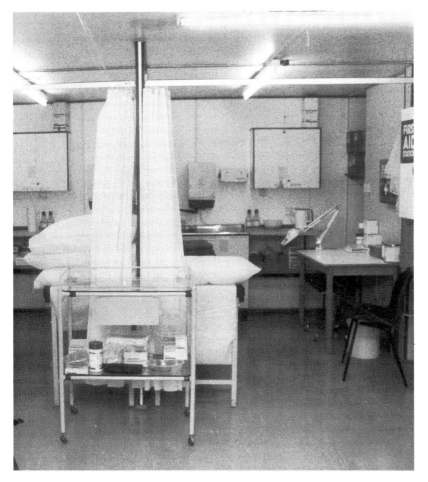

Fig. 11.3. A first-aid room on a construction site (courtesy Bovis Construction Ltd)

- one sterile unmedicated dressing
- one sterile triangular bandage
- six safety pins.

Shared first-aid facilities

The regulations require employers to provide first-aid facilities for their employees. It follows that subcontractors, engineers, other professional firms or clients who have staff on site should also make arrangements for the first aid of their employees, especially where their places of work are remote or separate from that of the main contractor. Resident engineers' and subcontractors' site offices should normally have a first-aid kit and trained first aiders or appointed persons.

Alternatively, subcontractors and others may decide to share first-aid facilities. Two or more employers on a site may make an agreement whereby one of them provides all the necessary first-aid equipment and personnel. The agreement should be recorded in writing and a copy kept by each employer concerned. The employers not providing first aid must satisfy themselves that the employer who has undertaken the duty is carrying it out to a satisfactory standard. Each employer must inform his employees of the first-aid arrangement on site.

First aiders
Legal requirements for first aiders
Paragraph 3(2) of the regulations requires an employer to provide an adequate number of suitable people for rendering first aid to his employees if they are injured or become ill at work. For this purpose a person is not suitable unless he has undergone

- such training and has such qualifications as the Health and Safety Executive may approve
- such additional training as may be appropriate in the circumstances.

Where the first aider is temporarily absent it is sufficient for an appointed person to take charge of the situation relating to an injured employee and of the equipment and facilities provided. An appointed person is someone of reasonable intelligence who should be able to handle an emergency and summon help.

For most sites the contractor should ensure that at least one first aider is normally present when the number of employees at work is between 50 and 150 and there should be at least one additional first aider for every 150 employees. Where there is shift working the contractor should ensure that sufficient first aiders are appointed to provide adequate coverage for each shift.

Where there are fewer than 50 employees there is no statutory duty to have a first aider but the employer must then ensure that an appointed person is present to take charge of the situation if a serious injury or major illness occurs. For any construction site, however, it is always preferable to have a trained first aider available.

First-aid training
The training of first aiders and the subsequent qualifications given by the British Red Cross Society, St John Ambulance Association and St Andrews Ambulance Association are approved by the Health and Safety Executive. Details of the courses and the centres where they are held may be obtained from the British Red Cross Society in London, the St John Ambulance in London and the St Andrews Ambulance Association in Glasgow.

The statutory certificate course for first aiders takes four consecutive days and covers

- resuscitation
- control of bleeding

- treatment of shock
- recognition of illness
- treatment of injuries
- simple record-keeping
- poisoning
- treatment of the unconscious patient
- dressing and immobilisation of injured parts
- contents of first-aid boxes and their uses
- transport of sick and injured patients
- treatment of burns and scalds
- personal hygiene when dealing with wounds
- communications and delegation in an emergency.

The statutory first-aid certificate lasts for three years, after which a requalification course is necessary. However, one-day refresher courses are available for those who wish to update their training every year. This is also recommended for appointed persons.

One-day courses are available for appointed persons at a number of centres. The courses give training in the handling of emergencies and the priorities for the care of a casualty when first aid is being rendered. The syllabus covers

- priorities for first aid
- asphyxia
- resuscitation
- shock and blood loss
- burns and scalds
- treatment of the unconscious casualty
- treatment of heart attacks
- treatment of strokes
- common workplace injuries.

Selection of staff for first-aid training

In terms of the turnover of most companies involved in the construction industry, the cost of first-aid training is relatively small and it is expected that most firms will send many more employees for first-aid training than the statutory minimum. It is customary for first aiders to receive a small salary increment in recognition of their special duties and occasional personal time involved in reading technical literature and other activities relating to first aid. It is recommended that chartered engineers and builders, and especially those on construction sites, should attend a first-aiders course or at least the appointed person's course. Apart from providing relief of suffering to the injured, the knowledge helps those in managerial positions to stay calm and act in a positive and sympathetic manner when an accident occurs.

First aid

First aid is a post-accident strategy vital to prevent loss of life and future deterioration in health following accidental injury or sudden illness. First aid is defined as the skilled application of accepted principles of treatment on the occurrence of an accident or in the case of sudden illness

- to sustain life
- to prevent deterioration in an existing condition
- to promote recovery.

The most important areas of first-aid treatment are

- restoration of breathing (resuscitation)
- control of bleeding
- prevention of collapse.

The first aider should take care not to become a casualty him or herself, for example, by entering a dangerous atmosphere without breathing apparatus or by entering a collapsed excavation without protection to prevent further collapse. Protective clothing should be worn and safety equipment used to reduce the risk of further injuries. Where help or an ambulance is needed they should be sent for without delay.

Simple first-aid procedures

As many employees as possible should be trained in simple first-aid procedures, namely resuscitation, control of bleeding and treatment of the unconscious patient, so as to ensure that casualties receive prompt attention. A number of simple first-aid procedures are outline in the following paragraphs (acknowledgements to *RoSPA Health and safety practice* by Stranks and Dewis[2]).

Resuscitation. Resuscitation (artifical respiration) may be required where the victim has suffered electric shock, gassing, suffocation or drowning. In each case, certain preliminary action is necessary as follows.

- *Electric shock.* Switch off the current if possible, otherwise pull the victim from contact, using heavy-duty insulating gloves, a rubber sheet, piece of dried timber, cloth, a folded newspaper, rope, the victim's own clothing if dry or other dry non-conducting material. Extreme care must be taken not to touch the victim's skin before the current is switched off.
- *Gassing.* Remove the victim to the fresh air or gas-free atmosphere as quickly as possible, ensuring that the rescuer is wearing suitable respiratory protection (e.g. breathing apparatus).
- *Suffocation.* If the victim has been buried in loose materials (e.g. in a trench collapse), immediately and quickly clear any debris from his mouth and nose. It may be desirable in this case to begin mouth-to-mouth resuscitation immediately the victim's head has been uncovered.
- *Drowning.* Remove the victim from the water with all speed. Clear any debris from his mouth. If rescue is by boat begin resuscitation in the boat. When the victim is clear of danger, resuscitation should be started immediately as the first minutes are vital, and should be continued without interruption until breathing is restored or until a doctor certifies that life is extinct.

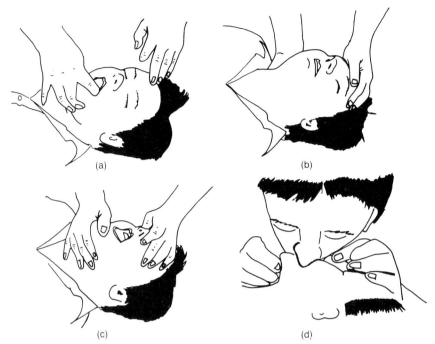

Fig. 11.4. Resuscitation procedure: (a) place the casualty on his back and clear his mouth; (b) tilt his head to open airway while supporting his jaw; (c) kneel beside casualty, open his mouth and pinch his nose; (d) open your mouth, take a deep breath, seal his mouth with yours and breathe firmly into it; when you see casualty's chest rise, remove your mouth and let his chest fall; continue until the casualty is breathing

Resuscitation procedure is described in Fig. 11.4.

Bleeding. Where bleeding is more than minimal, control it by direct pressure. Apply a pad of sterilised dressing or if necessary direct pressure with the fingers or thumb on the bleeding point. Raising a limb if the bleeding is sited there will help to reduce the flow of blood (unless the limb is fractured).

Unconsciousness. Where the patient is unconscious, care must be taken to keep the airways open. This is done by clearing the mouth and ensuring that the tongue does not block the back of the throat. Where possible the patient should be placed in the recovery position (*see* Fig. 11.5).

Broken bones. Unless he is in a position which exposes him from further danger, no attempt should be made to move a casualty with suspected broken bones or injured joints until the injured parts have been supported so that they cannot move or be moved separately (from the rest of the body).

Burns and scalds. Burns and scalds should be treated by flushing the affected area with plenty of clean cool water before applying a sterilised dressing or a clean towel. Where the burn is large or deep, a sterilised

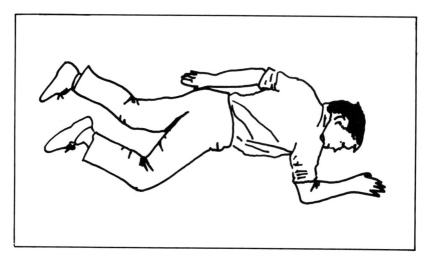

Fig. 11.5. Put the casualty in the recovery position

dressing should be applied. It is most important that blisters do not burst and that no attempt is made to remove clothing which may be sticking to burns or scalds. In the case of chemical burns any contaminated clothing which shows no signs of sticking to the skin should be removed, and all affected parts of the body flushed with plenty of cool clean water, ensuring that all the chemical substance is so diluted as to be rendered harmless. A sterilised dressing should be applied to exposed damaged skin and clean towels to damaged areas where the clothing cannot be removed. Care should be taken when treating such a casualty to avoid personal contamination.

Foreign bodies in the eye. Unless the object can be removed easily with a clean piece of moist material, no attempt should be made to remove the object. The eye should be irrigated with clean cool water. People with eye injuries or foreign bodies in the eye which cannot be removed in simple fashion should always be sent to hospital immediately, with the eye covered by a sterilised eye-pad dressing. Note: It is now well-established medical opinion that no first aiders should attempt to remove any form of foreign body from the eye of an injured person.

Chemical substances in the eye. The open eye should be flushed at once with cool clean water, the treatment continuing for five to ten minutes or even longer. If the contamination is more than minimal the casualty should be taken to hospital.

AIDS. Although the risk of first aiders contracting AIDS or other communicable diseases while treating patients is extremely small, recommendations are published by the Health and Safety Executive in the Construction Industry Advisory Committee Health hazard information sheet No. 6 AIDS (*see* Appendix 4). First aiders should cover their own cuts or abrasions with waterproof dressings before treating a patient. Wounds should be thoroughly irrigated and washed with soap and water before they are dressed. The first aider should always wash his or her hands after treating a wound. Although no cases of AIDS have been reported anywhere by transmission from saliva, a portable mouth-to-mouth resuscitation device incorporating a one-way valve can be used.

Reporting injuries
Reporting an injury or dangerous occurrence

The Reporting of Injuries, Diseases and Dangerous Occurrences Regulations 1985, often referred to as RIDDOR, require the notification in writing to the enforcing authority of all injuries resulting from accidents at work which cause incapacity for more than three days. In addition, any of the following events must first be notified to the enforcing authority by the quickest practicable means

- the death of any person as a result of an accident in connection with work
- any person suffering any of the following injuries or conditions as a result of an accident in connection with work; these are otherwise referred to as specified major injuries
  o fracture of the skull, spine or pelvis
  o fracture of any bone in the arm or wrist or leg or ankle
  o amputation of hand, foot, finger, thumb or toe
  o injury requiring immediate medical treatment resulting from electric shock
  o loss of consciousness resulting from lack of oxygen
  o decompression sickness
  o acute illness requiring treatment resulting from absorption of any substance by inhalation, ingestion or through the skin
  o acute illness requiring medical treatment where there is reason to believe that this resulted from exposure to a pathogen or infected material
  o any other injury which results in the person injured being admitted immediately into hospital for more than 24 hours.
- any of the dangerous occurrences listed in Appendix 6.

In the case of a death, a specified major injury or dangerous occurrence notification must be by telephone. This must be followed up within seven days by a written report to the enforcing authority on form F2508, available from HMSO. The enforcing authority is the body responsible for enforcing the HSW Act in the premises where or in connection with the work at which the reportable injury or dangerous occurrence happened. It may thus be either the Health and Safety Executive or the local authority. Construction site injuries and dangerous occurrences should be reported to the local office of the Health and Safety Executive.

Keeping records

A record must be made and kept of all reportable injuries and dangerous occurrences. The record must contain in each case the following information

- the date and time of the accident causing the injury or dangerous occurrence
- the following particulars about the person affected
  - full name
  - occupation
  - nature of injury or condition
- the place where the accident or dangerous occurrence happened
- a brief description of the circumstances.

Reporting a disease

Diseases are reported on form F 2508A. A leaflet *Reporting a case of disease* (HSE 17), available from the Health and Safety Executive (*see* Appendix 4), provides a brief guide to the reporting and record-keeping of cases of disease. The new regulations came into force in 1986.

Under the regulations, certain listed diseases must be reported either by employers or by the self-employed when those diseases are linked to specified types of work. For example, asbestosis should be reported where the person suffering has a job working with or handling asbestos. Leaflet HSE 17 gives a comprehensive list of reportable diseases and linked work activities. The reports are normally sent to the local office of the Health and Safety Executive or, in the case of offices, shops and restaurants, to the Environmental Health Department of the local authority.

Diseases related to construction work should always be reported to the HSE.

Internal organisation for reporting

It will be necessary for many of the larger companies in the construction industry to issue instructions for the reporting of injuries and dangerous occurrences to the enforcing authority. The company itself will need to define which person completes form F2508 or F2508A and whose responsibility it is to initiate the investigation, inform relatives and friends and deal with other administrative matters.

Accident investigation

The causes of accidents should be investigated

- to identify the cause and then to initiate measures to prevent a recurrence
- to gather information for use in any criminal or civil proceedings
- to confirm or refute a claim for industrial injury benefit
- to prepare notifications to be made to the enforcing authority.

It is vital that all accidents are investigated as quickly as possible. The longer the delay the less likely it is that the true facts will be ascertained. It is seldom that the cause of an accident is simple and it is equally seldom that only one or two people are to blame. Usually an accident arises from a number of acts or omissions by a number of people and the investigation should therefore be conducted in a calm, orderly atmosphere that encourages the witnesses to talk freely.

Investigation procedure

It is often necessary to conduct two or more interviews with witnesses. The first investigation will attempt to determine

- what caused the accident
- who was involved
- when, where and why it occurred
- how it could have been prevented.

It may then be necessary to consider obtaining further information to establish whether employees were adequately trained and whether proper work systems were being followed and, if not, how they could be improved to prevent a recurrence.

The cause of an accident should never be classified as carelessness because the only remedy to a careless act is for more care to be taken in the future. It is essential to be specific and to define the act or omission that caused the accident.

References

Chapter 1
1. Health and Safety Executive. *Blackspot construction*. HMSO, London, 1988.

Chapter 4
1. TOMLINSON M.J. General safety in excavation. *Hazards in construction*. Institution of Civil Engineers, London, 1971, 31–39.
2. TIMBER RESEARCH AND DEVELOPMENT ASSOCIATION. *Timber in excavation*. TRADA, London, 1990.
3. BRITISH STANDARDS INSTITUTION. *Code of practice for access and working scaffolds and special scaffold structures in steel*. BSI, London, 1981, BS 5973.
4. BRITISH STANDARDS INSTITUTION. *Code of practice for falsework*. BSI, London, 1982, BS 5975.
5. BRITISH STANDARDS INSTITUTION. *Code of practice for the use of safety nets on construction works*. BSI, London, 1972, CP 93.
6. BRITISH STANDARDS INSTITUTION. *Code of practice for the safe use of cranes (mobile cranes, tower cranes and derrick cranes)*. BSI, London, 1972, CP 3010.
7. HEALTH AND SAFETY EXECUTIVE. *Blackspot construction*. HMSO, London, 1988.
8. BRITISH STANDARDS INSTITUTION. *Rollover protective structures on earth moving machinery. Specification for crawler wheel loaders and tractors, backhoe loaders, graders, tractor scrapers, articulated shear dumpers*. BSI, London, 1987, BS 5527: Part 1.
9. BRITISH STANDARDS INSTITUTION. *Specification for falling object protective structures on earth moving machinery*. BSI, London, 1985, BS 5526.
10. DEPARTMENT OF TRANSPORT AND COUNTY SURVEYORS SOCIETY. *Notes for guidance in relation to the implementation of the requirements of HSW Act so far as they affect personnel who are required to undertake work on motorways and trunk roads*. London, 1985.
11. DEPARTMENT OF TRANSPORT, SCOTTISH DEVELOPMENT OFFICE AND WELSH OFFICE. *Traffic signs manual*. HMSO, London, 1974.
12. DEPARTMENT OF TRANSPORT. *Traffic signs and safety measures for minor works on minor roads*. DTp, London, 1989, Advice note TA 6/80.
13. BRITISH STANDARDS INSTITUTION. *Safety in tunnelling in the construction industry*. BSI, London, BS 6164.
14. BRITISH STANDARDS INSTITUTION. *Specification for lifejackets*. BSI, London, 1981, BS 3595.
15. BRITISH STANDARDS INSTITUTION. *Code of practice for demolition*. BSI, London, 1982, BS 6187.
16. BRITISH STANDARDS INSTITUTION. *Code of practice for the safe use of explosives in the construction industry*. BSI, London, 1988, BS 5607.

17. HEALTH AND SAFETY EXECUTIVE. *Noise at work*. HMSO, London, 1989.
18. HEALTH AND SAFETY EXECUTIVE. *The protection of persons against ionising radiations arising from any work activity*. HMSO, London, 1985, Approved code of practice.

Chapter 6
1. WATER RESEARCH CENTRE. *Sewers for adoption: a design and construction guide for developers*. WRC, Henley, 1985.
2. BRITISH STANDARDS INSTITUTION. *Code of practice for demolition*. BSI, London, 1982, BS 6187.
3. CONSTRUCTION INDUSTRY RESEARCH AND INFORMATION ASSOCIATION. *Quality assurance in civil engineering*. CIRIA, London, 1985, CIRIA report 109.

Chapter 7
1. ABBOTT H. *Safer by design*. Design Council, London, 1987.

Chapter 10
1. HEALTH AND SAFETY EXECUTIVE. *Noise at work*. HMSO, London, 1989.
2. BRITISH STANDARDS INSTITUTION. *Specification for eye protectors for industrial and non-industrial use*. BSI, London, 1987, BS 2092.
3. BRITISH STANDARDS INSTITUTION. *Specification for eye, face and neck protection against non-ionising radiation arising during welding and similar operations*. BSI, London, 1982, BS 1542.
4. BRITISH STANDARDS INSTITUTION. *Specification for filters for use during welding and similar operations*. BSI, London, 1977, BS 679.
5. BRITISH STANDARDS INSTITUTION. *Specification for filtering facepiece dust protectors*. BSI, London, 1980, BS 6016.
6. BRITISH STANDARDS INSTITUTION. *Optical performance of high-visibility garments and accessories for use on the highway*. BSI, London, 1985, BS 6629.

Chapter 11
1. HEALTH AND SAFETY EXECUTIVE. *First aid at work. Health and Safety (First-Aid) Regulations 1981*. HMSO, London, 1990, Approved code of practice.
2. STRANKS J and DEWIS M. *RoSPA Health and safety practice*. Pitman, London, 1986.

Appendices

Appendix 1. A selection of acts and statutory instruments relevant to the construction industry
Factories Act 1961
Abrasive Wheels Regulations 1970
Construction (General Provisions) Regulations 1961
Construction (Lifting Operations) Regulations 1961
Construction (Health and Welfare) Regulations 1966
Construction (Working Places) Regulations 1966
Highly Flammable Liquids and Liquefied Petroleum Gases Regulations 1972
Protection of Eyes Regulations 1974
Woodworking Machines Regulations 1974
Work in Compressed Air Special Regulations 1958 and 1960
Control of Pollution Act 1974
Health and Safety at Work etc. Act 1974
Asbestos (Licensing Regulations) 1983
Construction (Head Protection) Regulations. 1989
Construction (Management and Miscellaneous Duties) Regulations
  (in preparation 1990)
Control of Asbestos at Work Regulations 1987
Control of Lead at Work Regulations 1980
Control of Substances Hazardous to Health Regulations 1988
Diving Operations at Work Regulations 1981
Electricity at Work Regulations 1990
Health and Safety (First Aid) Regulations 1981
Ionising Radiations Regulations 1985
Noise at Work Regulations 1989
Reporting of Injuries, Diseases and Dangerous Occurrences Regulations 1985
Highways Act 1980
Mines and Quarries Act 1954

Appendix 2. Building Regulations 1985 made under the Building Act 1984

Approved document	Subject covered
A	Structure Loading and ground movement Disproportionate collapse
B	Fire spread
C	Site preparation and resistance to moisture
D	Toxic substances: cavity insulation
E	Sound: airborne and impact sound
F	Ventilation
G	Hygiene
H	Drainage and waste disposal
J	Heat producing appliances
K	Stairways, ramps and guards
L	Conservation of fuel and power
M	Access for disabled people

Appendix 3. Health and Safety Executive addresses

Headquarters
Health and Safety Commission, Baynards House, 1 Chepstow Place, Westbourne Grove, London W2 4TF (Tel: 071-243 6000).

Area offices

Area	Address	Tel/Fax	Local authorities covered
South West	InterCity House Mitchell Lane Victoria Street Bristol BS1 6AN	Tel: 0272 290681 Fax: 0272 262998	Avon, Cornwall, Devon, Gloucestershire, Somerset, Isles of Scilly
South	Priestley House Priestley Road Basingstoke RG24 9NW	Tel: 0256 473181 Fax: 0256 51744	Berkshire, Dorset, Hampshire, Isle of Wight, Wiltshire
South East	3 East Grinstead House, London Road East Grinstead West Sussex RH19 1RR	Tel: 0342 326992 Fax: 0342 312917	Kent, Surrey, East Sussex, West Sussex
London North	Maritime House 1 Linton Road Barking Essex IG11 8HF	Tel: 081-594 5522 Fax: 081-591 5183	Barking and Dagenham, Barnet, Brent, Camden, Ealing, Enfield, Hackney, Haringey
	Chancel House Neasden Lane London NW10 2UD	Tel: 081-459 8855 Fax: 081-459 2131	Harrow, Havering Islington, Newham, Redbridge, Tower Hamlets, Waltham Forest
London South	1 Long Lane London SE1 4PG	Tel: 071-407 8911 Fax: 071-403 7058	Bexley, Bromley, City of London, Croydon, Greenwich, Hammersmith & Fulham, Hillingdon, Hounslow, Kensington & Chelsea, Kingston, Merton, Richmond, Southwark, Sutton, Wandsworth, Westminster
East Anglia	39 Baddow Road Chelmsford CM2 OHL	Tel: 0245 284661 Fax: 0245 252633	Essex, Norfolk, Suffolk
Northern Home Counties	14 Cardiff Road Luton, Beds. LU1 1PP	Tel: 0582 34121 Fax: 0582 459775	Bedfordshire, Buckinghamshire, Cambridgeshire, Hertfordshire

Area	Address	Tel/Fax	Local authorities covered
East Midlands	5th Floor, Belgrave House, 1 Greyfriars, Northampton NN1 2BS	Tel: 0604 21233 Fax: 0604 30460	Leicestershire, Northamptonshire, Oxfordshire, Warwickshire
West Midlands	McLaren Building 2 Masshouse Circus Queensway Birmingham B4 7NP	Tel: 021-200 2299 Fax: 021-233 2176	West Midlands
Wales	Brunel House 2 Fitzalan Road Cardiff CF2 1SH	Tel: 0222 473777 Fax: 0222 473642	Clwyd, Dyfed, Gwent, Gwynedd, Mid Glamorgan, Powys, South Glamorgan, West Glamorgan
Marches	The Marches House Midway, Newcastle under Lyme, Staffs ST5 1DT	Tel: 0782 717181 Fax: 0782 620612	Hereford and Worcester, Shropshire, Staffordshire
North Midlands	Birbeck House Trinity Square Nottingham NG1 4AU	Tel: 0602 470712 Fax: 0602 411577	Derbyshire, Lincolnshire, Nottinghamshire
South Yorkshire	Sovereign House 40 Silver Street Sheffield S1 2ES	Tel: 0742 739081 Fax: 0742 755746	Humberside, South Yorkshire
West and North Yorkshire	8 St Pauls Street Leeds LS1 2LE	Tel: 0532 446191 Fax: 0532 450626	North Yorkshire, West Yorkshire
Greater Manchester	Quay House Quay Street Manchester M3 3JB	Tel: 061-831 7111 Fax: 061-831 7169	Greater Manchester
Merseyside	The Triad Stanley Road Bootle Merseyside	Tel: 051-922 7211 Fax: 051-922 5031	Cheshire, Merseyside
North West	Victoria House Ormskirk Road Preston PR1 1HH	Tel: 0772 59321 Fax: 0772 21807	Cumbria, Lancashire

Area	Address	Tel/Fax	Local authorities covered
North East	Arden House Regent Centre Regent Farm Road Gosforth Newcastle upon Tyne NE3 3JN	Tel: 091-284 8448 Fax: 091-285 9682	Cleveland, Durham, Northumberland, Tyne and Wear
Scotland East	Belford House 59 Belford Road Edinburgh EH4 3UE	Tel: 031-225 1313 Fax: 031-225 6783	Borders, Central, Fife, Grampian, Highland, Lothian, Tayside and the island areas of Orkney and Shetland
Scotland	Royal Exchange Assurance House 314 St Vincent Street Glasgow G3 8XG	Tel: 041-204 2646 Fax: 041-248 2760	Dumfries and Galloway, Strathclyde and the Western Isles

Health and Safety Executive public enquiry points
HSE Library and Information Service, Broad Lane, Sheffield S3 7HQ (Tel: 0742 752539).

HSE Library and Information Service, St Hugh's House, Trinity Road, Bootle, Merseyside L20 3QY (Tel: 051-951 4381).

HSE Library and Information Service, Baynards House, 1 Chepstow Place, Westbourne Grove, London W2 4TF (Tel: 071-221 0870).

Source: Health and Safety Executive, 1989.

Appendix 4. HSC/E publications relevant to construction

Abbreviations

PM	Guidance Note, plant and machinery
GS	Guidance Note, general series
EH	Guidance Note, environmental hygiene
MS	Guidance Note, medical
CS	Guidance Note, chemical safety
SS	Summary sheet for small contractors
HS(G)	Health and Safety guidance
HS(R)	Health and Safety regulations
IAC	Industry Advisory Committee

Abrasive wheels (*see* Power tools)

Access

Access to tower cranes HMSO 1979	PM9
Construction hoists HMSO 1981	PM27
General access scaffolds HSE 1987	SS3
General access scaffolds HMSO 1982	GS15
Safe use of ladders HSE 1988	SS2
Safe use of ladders, step ladders and trestles HMSO 1984	GS31
Safety at rack and pinion hoists HMSO 1981	PM24
Safety in working with power operated mobile work platforms HMSO 1982	HS(G)19
Suspended access equipment HMSO 1983	PM30
Suspended cradles HSE 1988	SS5
Tower scaffolds HSE 1987	SS10

Accidents

Fatal accidents in construction HMSO 1978 (ISBN 0 11 883419 3)	
Guidance on the collection and use of accident information in the construction industry HMSO 1983 (ISBN 0 11 883727 3)	
Guide to the reporting of injuries, diseases and dangerous occurrences HMSO 1966	HS(R)23
Management's responsibilities in the safe operation of mobile cranes: report on three crane accidents HMSO 1980 (ISBN 0 11 883301 4)	
One hundred fatal accidents in construction HMSO 1978 (ISBN 0 11 883072 6)	
Report that accident HSE 1988	HSE 21
Reporting an injury or a dangerous occurrence HSE 1988	HSE 11 rev.

Adhesives

Petroleum based adhesives in building operations HMSO 1977	EH7

AIDS

AIDS   HSE 1987 (Health hazard information sheet No. 6)

Alcohol and problem drinkers

The problem drinker at work   HMSO 1981   (HSE occasional paper OP1)

Asbestos

Alternatives to asbestos products:
a review   HMSO 1986 (ISBN 0 11 883812 1)
Asbestos: exposure limits and measurement of
airborne dust concentrations   HMSO 1988   GS10
Asbestos and you   HSE 1984 (IND(G)17(L))
The control of asbestos at work regulations 1987
HMSO 1988   Approved code of practice COP 21
The control of asbestos dust during work with asbestos —
asbestos removal techniques (in preparation 1988)
Enclosures provided for work with asbestos
insulation and coatings (in preparation 1988)
A guide to the asbestos (licensing) regulations 1983
HMSO 1984   HS(R)19
Probable asbestos dust concentrations
at construction processes   HMSO 1984   EH35
The provision, use and maintenance of hygiene
facilities for work with asbestos
insulation and coatings   HMSO 1986   EH47
Respiratory protective equipment for use
against asbestos   HMSO 1985   EH41
Training operatives and supervisors for work
with asbestos insulation and coatings   HMSO 1988   EH50
Work with asbestos   HSE 1987   SS9
Work with asbestos cement   HMSO 1984   EH36
Work with asbestos insulating board   HMSO 1984   EH37
Work with asbestos insulation, asbestos coating
and asbestos insulating board   HMSO 1988
Approved code of practice COP 3
Working with asbestos in the asbestos removal industries —
a guide for supervisors and safety representatives
(in preparation 1988)
Working with asbestos in the construction industry — a guide
for supervisors and safety representatives (in preparation 1988)

Cement

Cement HSE 1985 (Health hazard information sheet No. 1)
Work with asbestos cement   HMSO 1984   EH36

Chain saws

Chain saws   HMSO 1982   PM31

Check lists

Construction site safety checklist   HSE 1988

Children

Accidents to children on construction sites   HMSO 1977   GS7
Building sites bite — stay away (Poster)   HSE 1988 (IND(G)52(P))

Confined spaces

Entry into confined spaces   HMSO 1977   GS5
Entry into confined spaces   HSE 1988   SS15
Safe use of propane and other LPG cylinders   HSE 1988   SS11

Demolition
   Flame cutting and welding with compressed gases HSE 1988   SS12
   Health and safety in demolition work
      Part 1 preparation and planning HMSO 1984   GS29/1
      Part 2 legislation HMSO 1984   GS29/2
      Part 3 techniques HMSO 1984   GS29/3
      Part 4 health hazards HMSO 1985   GS29/4
   Hot work: welding and cutting on plant
      containing flammable materials   HS(G)5
   Hot work on tanks and drums HSE 1985 (IND(G)35(L))
   Safety in demolition work HMSO 1979 (ISBN 0 11 883242 5)

Diseases
   Guide to the reporting of injuries, diseases and dangerous
     occurrences regulations 1985 HMSO 1986 (ISBN 0 11 883858 X)
   Reporting a case of disease: a brief guide to RIDDOR 1985 HSE 1986
                                                                   HSE17

Diving
   A guide to diving operations at work regulations 1981 HMSO 1981 HS(R)8

Dumper trucks
   Safe working with small dumpers HMSO 1983 (ISBN 0 11 883693 5)

Dust
   Control of hardwood dust HSE 1987 (IND(S)21(C))
   Dust in the workplace: general principles of protection HMSO 1984 EH44
   Wood dust: hazards and precautions a guide for employers
   HSE 1987 (IND(S)10(L))
   (*see also* Asbestos)

Electric cables
   Avoiding danger from buried services HSE 1987   SS7
   Avoiding danger from buried electricity cables HMSO 1985   GS33
   Buried cables: beware HSE 1985 (IND(G)30(L))

Electric shock
   Protection against electric shock HMSO 1984   GS27

Electricity
   Avoidance of danger from overhead electrical lines HMSO 1980   GS6
   Electrical safety in arc welding HMSO 1986   PM64
   Electrical test equipment for use by electricians HMSO 1986   GS38
   Electricity on construction sites HMSO 1983   GS24
   Flexible leads, plugs and sockets HMSO 1985   GS37
   The safe use of portable electrical apparatus HMSO 1983   PM32
   Selection and use of electric headlamps HMSO 1984   PM38
   The use of portable electric equipment on construction sites HSE 1987 SS6

Excavations
   Safety in excavations HSE 1987   SS8

242 CONSTRUCTION SAFETY HANDBOOK

Falsework
    Checklist for supervisors and charge hands erecting falsework
        HSE 1987 (IND(G)44(L))
    Safety in falsework for insitu beams and slabs
        HMSO 1987   HS(G)32

Fire
Fire precautions in pressurised workings HMSO 1983	GS20
Flame cutting and welding with compressed gases HSE 1988	SS12
Industrial use of flammable gas detectors HMSO 1979	CS1
Petroleum based adhesives in building operations HMSO 1977	EH7
Safe use of propane and other LPG cylinders HSE 1988	SS11
Spraying of highly flammable liquids HMSO 1977	EH9

First aid
    First aid at work   HMSO 1981   HS(R)11
    First aid at work: general guidance for inclusion
       in first aid boxes   HMSO 1987 (IND(G)4rev(P))
    Health and Safety (First Aid) Regulations 1981: Approved
       Code of Practice No. 4   HMSO 1990 (ISBN 0 11 885536 0)

Flammable liquids
Cleaning and gas freeing of tanks containing flammable residues HMSO 1985	CS15
Highly flammable materials on construction sites HSE 1988	HS(G)3
Petroleum based adhesives in building operations HMSO 1977	EH7
Spraying of highly flammable liquids HMSO 1977	EH9

Fork lift trucks
Safety in working with fork lift trucks HMSO 1979	HS(G)6
Working platforms on fork lift trucks HMSO 1981	PM28

Health and safety
    Articles and substances used at work   HSE 1988 (IND(G)1(L)rev)

Health hazards
    AIDS   HSE 1987 Health hazard information sheet   No. 6
    Cement   HSE 1985 Health hazard information sheet   No. 1
    Cold weather   HSE 1985   (Health hazard information sheet)   No.2
    Lead   HSE 1987   (Health hazard information sheet)   No.4
    Noise   HSE 1986   (Health hazard information sheet)   No.3
    Solvents   HSE 1987   (Health hazard information sheet)   No.5
    Skin hazards   HSE 1987   (Health hazard information sheet)   No.7

Isocyanates
    Isocyanates: toxic hazards and precautions   HMSO 1984   EH16

Ladders (*see* Access)

Lead
    Control of Lead: air sampling techniques and strategies   HMSO 1981   EH28

Control of lead at work: approved code of practice
  HMSO 1985 (ISBN 0 11 883780X)
Lead   HSE 1985 (Health hazard information sheet No.4)
Lead and you   HSE 1986   MS(A)1

Legislation
  Articles and substances used at work   HSE 1988 (IND(G)1(L)rev)
  General legal requirements   HSE 1988   SS1
  Master list of health and safety at work legislation and
    approved codes of practice
    Part I (Post HSW Act)   HSE 1987 (RPD A1/118/1984)
    Part II (Pre HSW Act)   HSE 1986 (RPD A1/118/1984)

Lifting appliances
  Access to tower cranes   HMSO 1979   PM9
  Cable laid slings and grommets   HMSO 1987   PM20
  Construction goods hoists   HSE 1988   SS13
  Construction hoists   HMSO 1981   PM27
  The crane rig   HSE 1986
    (Technical information leaflet engineering and metallurgy 10)
  Erection and dismantling of tower cranes   HMSO 1976   PM3
  Excavators used as cranes   HMSO 1984   PM42
  Inclined hoists used in building and construction work   HMSO 1987   PM63
  Lifting gear standards   HMSO 1985   PM54
  Safe working with overhead travelling cranes   HMSO 1985   PM55
  Safety at power operated mast work platforms   HMSO 1985   HS(G)23
  Safety at rack and pinion hoists   HMSO 1981   PM24
  Scotch derrick cranes   HMSO 1984   PM43
  Small lifting appliances   HSE 1988   SS14
  Specification for automatic safe load indicators   HSE 1985
    (ISBN 0 7176 0204 4)
  Suspended cradles and small lifting appliances   HSE 1987   SS5
  Training of crane drivers and slingers   HMSO 1986   GS39
  Wedge and socket anchorage for wire ropes   HMSO 1985   PM46

Liquefied petroleum gas
  Flame cutting and welding with compressed gases   HSE 1988   SS12
  Safe use of propane and other LPG cylinders   HSE 1987   SS11
  Storage of LPG at fixed installations   HMSO 1987   HS(G)34
  The storage and use of LPG on construction sites   HMSO 1981   CS6

Loading and unloading
  Cable laid slings and grommets   HMSO 1987   PM20
  Safety in the stacking of materials   HMSO 1971   HSW47
  Safety in the use of timber pallets   HMSO 1978   PM15

Maintenance
  Deadly maintenance: plant and machinery: a study of
    accidents at work   HMSO 1985 (ISBN 0 11 883805 9)
  Deadly maintenance: roofs: a study of accidents at work
    HMSO 1985 (ISBN 0 11 883804 0)

Management
 Managing health and safety in construction: main contractor/
   sub contractor projects Pt I   HSC Coniac 1987 (ISBN 0 11 883918 7)
 Managing health and safety in construction:
   management contracting Pt II   HSC Coniac 1988 (ISBN 0 11 883939 6)

Mineral wool
 Exposure to mineral wools   HMSO 1986   EH46

Noise
 It's your hearing, protect it. Noise alert for construction
   workers   HSE 1986   IAC/L17
 Level of training for technicians making noise surveys   HMSO 1977   EH14
 Noise   HSE 1986 (Health hazard information sheet No.3)
 Noise at woodworking machines   HSE 1988 (IND(S)22(L))
 Noise from portable breakers   HSE 1986   IAC/L21
 Noise in construction: guidance on noise control and
   hearing conservation measures   HMSO 1986 (ISBN 0 11 883877 6)
 Noise from pneumatic systems   HMSO 1985   PM56

Occupational health
 Review your occupational health needs: employers guide
   HMSO 1988 (IND(G)57(L))

Pitch and tar
 Skin cancer caused by pitch and tar   HSE 1984   MS(B)4

Power tools
 Safety in the use of abrasive wheels   HMSO 1984   HS(G)17
 Training advice on the mounting of abrasive wheels   HMSO 1983   PM22
 Chain saws   HMSO 1982   PM31
 Guarding of portable pipe-threading machines   HMSO 1984   PM1
 Pneumatic nailing and stapling tools   HMSO 1979   PM17
 Safety in the use of cartridge operated tools   HMSO 1978   PM14
 Safety in the use of woodworking machines   HMSO 1981   PM21
 The use of portable electric tools on construction sites   HSE 1987   SS6

Pressure testing
 Safety in pressure testing   HMSO 1976   GS4

Pressurised workings
 Fire precautions in pressurised workings   HMSO 1983   GS20

Protective clothing and equipment
 Protective clothing and footwear in the construction industry
   HSC 1986   IAC/L16
 (For protective equipment for use against asbestos *see* Asbestos)

Roofwork
 Deadly maintenance: roofs (*see* Maintenance)
 Safety in roofwork   HSE 1987   SS4
 Safety in roofwork   HMSO 1987   HS(G)33

Scaffolds (*see* Access)

Skin diseases
Occupational skin diseases HMSO 1981	EH26
Save your skin: advice to employers HSE 1987	MS(B)9
Save your skin: occupational contact dermatitis HSE 1987	MS(B)6 rev
Skin cancer caused by pitch and tar HSE 1984	MS(B)4

Skin hazards   HSE 1987 (Health hazard information sheet no. 7)

Structures
Safe erection of structures
Pt 1 initial planning and design HMSO 1984	GS28/1
Pt 2 site management and procedures HMSO 1985	GS28/2
Pt 3 working places and access HMSO 1986	GS28/3
Pt 4 legislation and training HMSO 1986	GS28/4

Toxic substances
Monitoring strategies for toxic substances HMSO 1984	EH42
Toxic substances: a precautionary policy HMSO 1977	EH18

Transport
Danger: Transport at Work   HSE 1985 (IND(G)22(L))
Safe working with small dumpers   HMSO 1983 (ISBN 0 11 883693 5)

Underground services
Avoiding danger from buried services   HSE 1987   SS7
(*See also* Electric cables)

Welding
Electrical safety in arc welding HMSO 1986	PM64
Flame cutting and welding with compressed gases HSE 1988	SS12
Hot work: welding and cutting on plant containing flammable materials HSE 1979	HS(G)5
Welding HMSO 1978	MS15

These publications may be purchased from the Health and Safety Executive public enquiry points listed in Appendix 3.

Source: Health and Safety Executive, 1989.

Appendix 5. Contents of first-aid boxes and travelling first-aid kits

Item	First-aid boxes					Travelling first-aid kits
	Numbers of employees					
	1–5	6–10	11–50	51–100	101–150	
Guidance card or leaflet	1	1	1	1	1	—
Individually wrapped sterile adhesive dressings	10	20	40	40	40	6
Sterile eye pads, with attachment: an example of a suitable eye pad currently available would be the Standard Dressing No. 16 BPC	1	2	4	6	8	—
Triangular bandages (if possible, sterile)	1	2	4	6	8	1
Sterile coverings for serious wounds (if triangular bandages not sterile)	1	2	4	6	8	1
Safety pins	6	6	12	12	12	6
Medium sized sterile unmedicated dressings approx. 10 cm × 8 cm: examples of suitable dressings currently available are the Standard Dressings No. 8 and No. 13 BPC	3	6	8	10	12	1
Large sterile unmedicated dressings approx. 13 cm × 9 cm: examples of suitable dressings currently available are the Standard Dressings No. 9 and No. 14 BPC and the Ambulance Dressing No. 1	1	2	4	6	10	—
Extra large sterile unmedicated dressings approx 28 cm × 17.5 cm: an example of a suitable dressing currently available would be the Ambulance Dressing No. 3	1	2	4	6	8	—
If tap water is not available, sterile water or sterile normal saline in disposable containers, each holding at least 300 ml for eye irrigation, needs to be kept near the first-aid box. At least these numbers of containers should be kept	1	1	3	6	6	—

*Source: Health and Safety Executive IND(a)3(L).

Appendix 6. Dangerous occurrences to be notified (abridged)
For complete list see HSE II (Rev) *Reporting an injury or a dangerous occurrence.*

1. The collapse of, the overturning of, or the failure of any load-bearing part of
   - any lift, hoist, crane, derrick or mobile powered access platform;
   - any excavator;
   - any pile driving frame or rig having an overall height, when operating, of more than 7 m

2. The following incidents at a fun fair while the relevant device is in use or under test
   - the failure of any load-bearing part of any amusement device which is designed to allow passengers to move or ride on it;
   - the failure of any safety arrangement connected with such a device

3. Explosion, collapse or bursting of any closed vessel, in which the internal pressure was above or below atmospheric pressure

4. Electrical short circuit or overload attended by fire or explosion which resulted in the stoppage of the plant involved for more than 24 hours

5. An explosion or fire occurring in any plant or place which resulted in the suspension of normal work for more than 24 hours

6. The sudden, uncontrolled release of 1 tonne or more of highly flammable liquid, gas or flammable liquid above its boiling point

7. A collapse or partial collapse of any scaffold which is more than 5 m high and where the scaffold is slung or suspended, a collapse or partial collapse of the suspension arrangements which causes a working platform or cradle to fall more than 5 m

8. Any unintended collapse or partial collapse of
   - any building or structure under construction, reconstruction, alteration or demolition, or of any falsework, involving a fall of more than 5 tonnes of material
   - any floor or wall of any building being used as a place of work

9. The uncontrolled or accidental release or the escape of any substance or pathogen from any apparatus, pipe-line, storage vessel, landfill site, or exploratory land drilling site

10. Any ignition or explosion of explosives, where the ignition or explosion was note intentional

11. Failure of any freight container or failure of any load-bearing part thereof

12. Either of the following incidents in relation to a pipe-line as defined by Section 65 of the Pipe-Lines Act 1962
    - the bursting, explosion or collapse of a pipe-line or any part thereof,
    - the unintentional ignition of anything in a pipe-line, or of anything which immediately before it was ignited was in a pipe-line

13. Any incident

    - in which a road tanker or tank container used for conveying a dangerous substance by road
      - overturns;
      - suffers serious damage to the tank;
    - in which there is, in relation to such a road tanker or tank container
      - an uncontrolled release or escape of the dangerous substance being conveyed
      - a fire which involves the dangerous substance being conveyed

14. Any incident involving a vehicle conveying a dangerous substance by road, other than a vehicle to which paragraph 13 applies, where there is

    - an uncontrolled release or escape of the dangerous substance being conveyed from any package or container
    - a fire which involves the dangerous substance being conveyed

15. Any incident where breathing apparatus malfunctions in such a way as to be likely either to deprive the wearer of oxygen or to expose the wearer to contaminant

16. Any incident in which plant or equipment either comes into contact with an uninsulated overhead electric line in which the voltage exceeds 200 volts, or causes an electrical discharge from such an electric line

17. Any case of an accidental collision between a locomotive or a train and any other vehicle at a factory or dock.

Index

Abbeystead explosion, 26
Accidents and first aid, 218–231
  AIDS, 228–229
  first aid, 225–229
  first aiders, 224–225
  first-aid facilities, 220–224
  investigation, 230–231
  keeping records, 229–230
  organisation for reporting, 230
  reporting diseases, 230
  reporting injuries, 229
  site arrangements, 218–220
Accident statistics, 3–7
  causes of accidents, 7–9
  economic effect of accidents, 6
  fatal and major injuries, 4, 5, 8, 9
  reasons for the poor record, 6–7
  reliability of the statistics, 9–10
Approved codes of practice, 18
Association of Consulting Engineers, 22, 24
Atkins, W. S., 206

Basements, 47–48
Breathing apparatus, 116–117, 134, 213–214
British Red Cross Society, 207
Brooklands School of Management, 206
Building Employers Confederation
  publications, 201
  training, 206, 207
Building Regulations, 18, 235

Checking designs and drawings, 172–173
Chemical hazards, 10, 132–135
  contact with the skin, 132
  ingestion, 135
  inhalation, 132–135
  asbestos dust, 132–133
  cadmium, 134
  carbon dioxide, 133–134
  carbon monoxide, 133
  hydrogen sulphide, 134
  lead, 134–135
  nitrous fumes, 134
  silica dust, 133
  solvents, 135
  welding fumes, 135
  zinc, 135
Civil law, 21–27
  law of contract, 21–25, see also Contract law
  liquidated damages, 21
  precedent, 21
  tort, 21, 25–26, see also Tort
  unliquidated damages, 21
Competent persons, 184, 186
Confined spaces, 49, 106–107, 112, 115–117
  safe systems of work in, 116
  working precautions in, 117
CONIAC, 37, 39
Construction Industry Training Board, 206
Construction Management Regulations, 19, 162
Construction manager, duties of, 153
Construction Regulations, 13–15
  obligations under, 14–15
Construction safety manual, 201
Contaminated sites, 117–118
  identification and remedial action, 118
  site precautions, 118
Contract law, 22–25
  conditions of contract, 22

engineering contracts, 24–25
failures involving injury, 25
ICE Conditions of Contract, 22–24
limitation, 24–25
Control of Industrial Major Accident Hazards (CIMAH) Regulations, 195
Control of Substances Hazardous to Health (COSHH) Regulations, 20
Cranes, 86–89
  hazards and safety measures, 87–91
    dropping the load, 89
    electrocution, 89–91
    erection and dismantling, 91
    overturning and structural failure, 87–89
  safe use of, 86–88
    automatic safe load and load radius indicators, 97–98
    piling operations, 93
    planning lifting operations, 86–87
    selection of operatives, 87
    tandem lifting, 82
    testing and examination, 97
    trapping, 91
  types of, 92–97
    guyed derricks, 97
    mobile cranes, 92–93
    scotch derricks, 96
    tower cranes, 94–96
Criminal law, 11–21

Demolition, 121–131
  explosives, 127–129
  hand, 123–124
  health hazards, 129–131
  inspection, 121
  mechanical, 125–127
  method statement, 122
  supervision, 122
Disobeying safety orders, 27

Economic effect of accidents, 6
Emergency procedures plan, 219–220
Employer
  and employee, 26–27
    disobeying safety orders, 27
    vicarious liability, 27
  definition of, 140
  duties of management, 147–151
  policy statements, 140–146
  safety organisation, 146–157

Enforcement of health and safety law, 28–39
Engineering contracts, 24–25
  care and diligence, 24
Engineer's safety adviser, 160
English law, 11–27
Erection of structural frameworks, *see* Structural frameworks
Excavations, 40–51
  basements, 47–48
  deep bored piles, 50–51
  shafts, 48–50
  trenches, 40–46
    causes of collapse, 41–43
    prevention of collapse, 43–45
Eye protection, 213

Factories Act, 12–13
Failure mode and effect analysis, 193–194
Failures involving injury, 25
Falsework, 67–75
  construction, 73–75
    adjustable steel props and forkheads, 74
    checking, 75
    foundations, 74
    loading, 74–75
    materials, 73
    scaffold support components, 74
    temporary works co-ordinator, 75
  design, 68–73
    buckling, 72
    design brief, 72–73
    forces on mobile falsework, 71
    foundations, 70
    lateral stability, 71
    loading, 68–70
    materials, 70–71
    traffic access, 72
Fault-tree analysis, 193
Federation of Civil Engineering Contractors, 22, 202, 206
Films on safety topics, 203
First aid
  facilities, 220–224
    first-aid boxes, 222–223
    first aiders, 224–225
    first-aid rooms, 222
    legal requirements, 220, 222
    selection of staff, 226
    self-employed people, 222
    shared facilities, 223–224

INDEX

training, 224–225
procedures, 225–229
  AIDS, 228–229
  bleeding, 227
  broken bones, 227
  burns and scalds, 227–228
  chemical substances in the eye, 228
  foreign bodies in the eye, 228
  resuscitation, 226–227
  unconsciousness, 227
Footwear, safety, 212

Gases and vapours, 131–135
Guidance notes, 200–201, 239–245

Health and safety
  legislation, history, 11–21
  organisation, 146–161
    consulting engineer, 159–161
    contractor, 152–154
    major employer, 152
    management contracting, 154–157
    safety groups, 157–158
    safety representatives and committees, 158–159
    subcontracting, 157
  policies, 140
    consulting engineer, 145–146
    contractor, 143–145
    major employer, 142–143
Health and Safety Agency, N. Ireland, 28
Health and Safety at Work etc. Act 1974, 15–21
  approved codes of practice, 18
  Construction Management Regulations, 19, 162
  delegated legislation, 11, 234
  HMS *Glasgow*, 20–21
  'reasonably practicable', 19–20
  substances hazardous to health, 20
Health and Safety Commission, 28
Health and Safety Executive, 28, 39, 236–238
  Factory Inspectorate, 32
  Guidance notes, 239–245
Health hazards, 10, 131–139
  biological, 114, 138–139
    Weil's disease (leptospiral jaundice), 112, 138–139
  chemical
    asbestos dust, 132–133
    cadmium, 134

carbon dioxide, 49, 106, 112, 116, 133–134
carbon monoxide, 49, 106, 116, 133
coal dust, 107
hydrogen sulphide, 106, 112, 134
lead, 134–135
nitrous fumes, 49–50, 106, 116, 134
silica dust, 106, 133
solvents, 135
welding fumes, 135
zinc, 135
physical, 135–138
  cold, 135–136
  compressed air, 110–111, 138
  heat, 136
  ionising radiations, 137–138
  noise, 136–137
  vibration, 137
Hearing protection, 212
Helmets, safety, 210
HM Factory Inspectorate, powers of, 32–34
  CONIAC, 37, 39
  Improvement Notice, 34
  Prohibition Notice, 35
  prosecution, 35
High-visibility garments, 214

Improvement Notice, 34, 36–37
Information about safety, 196–203
  Construction safety manual, 201–202
  films, 203
  for managers, 198–199
  guidance notes, 200–201, 239–245
  in-house, 196–199
  posters, 199–200
  practice notes, 199
  publications, 200–203
  safety handbooks, 197–198
Injuries, causes of major, 9
  annual totals, 3, 4
  rates per 100 000 employees, 5
Institution of Civil Engineers, 1
  Conditions of Contract, 22–24
  and HSW Act, 23–24
Institution of Occupational Safety and Health, 154, 206
Ionising radiation, 137–138

Ladders, 66–67
Latent Damage Act, 26
Legislation, health and safety, 11–21

Limitation
  in contract law, 24–25
  in tort, 25–26

Major accident hazard legislation, 195
Management
  competent person, 184, 186
  design and specification, 166–179
  management of construction, 180–189
  permits to work, 180, 184
  pre-contract activities, 162–164
  quality assurance
    construction, 186–190
    design, 174–179
  safety appraisal, 180–181
  safety audit, 179–180
  survey and investigation, 164–166
  systems for design offices, 172–176
  systems for safe construction, 162–190
  training, 205–206
Mandatory protective clothing and equipment, 208–209
Methane gas, 49, 50, 106, 116

Negligence, 25
  involving injury, 26

Occupation, fatal injuries by, 8
Outer wear, 214
Oxygen deficiency, 49, 106, 112, 113, 116, 134

Permits to work, 116, 180–181, 184
Physical hazards, 135–138
  cold, 135–136
  compressed air, 110–111, 138
  heat, 136
  ionising radiation, 137–138
  lasers, 138
  noise, 136–137
  vibration, 137
Posters, 199–200
Practice notes, 199
Prohibition Notice, 35, 38
Project manager, duties of, 146, 156–157, 159, 165, 199
Protective clothing, 208–217
  eye protection, 213
  hearing protection, 212
  high visibility garments, 214
  outer wear, 214
  respiratory protection, 213–214
  safety footwear, 212
  safety helmets, 210
  standards of, 209
  use on construction sites, 210–214
  wearing by professional staff, 209

Quality assurance
  construction, 186–189
    control of workmanship, 187
    flow sheets, 187–189
    non-compliance procedure, 189
    quality plan, 186–187
    review meetings, 189
  design, 174
    management manual, 174
    project quality plans, 176
  effectiveness of, 190

Reporting injuries, diseases and dangerous occurrences (RIDDOR), 229, 230
  internal organisation for, 230
Resident Engineer, duties of, 160–161
Respiratory protection, 213–214
Risk analysis, principles of, 191–195
Roadworks, 101–104
  safety of personnel, 102–103
  traffic safety at, 103
  underground services, 103–104
RoSPA Occupational Safety Training Centre, 207

Safety
  adviser, duties of, 153–154, 159–160
  and reliability, 191–195
    failure mode and effect analysis, 193–194
    fault-tree analysis, 193
    reliability, 194
  cases, 194–195
  director, duties of, 152–153
  equipment, 216–217
    footwear, 211–212
    helmets, 210
    safety belts and harnesses, 216–217
  groups, 157–158
  policies, 140–161
  representatives and safety committees, 158–159
  supervisors, training, 207
  training, 153–154, 159–160

Safety manual for mechanical plant construction, 202
St John Ambulance, 207, 224
Scaffolding, 51–67
  birdcage, 62–63
  causes of collapse, 57
  falls from, 53–56
  falls of materials, 56–57
  proprietary, 60–61
  putlog, 52
  slung, 64–65
  suspended, 65–66
  tied independent, 52–53
  tower, 61–62
  tube and fitting, 52–53
Self-employed persons, first aid, 222
Sewers, 111–115
  communications with workmen, 114–115
  hazards of work in, 111–112
  reducing hazards, 112–114
    dangerous atmospheres, 113–114
    falls, 112–113
    infection from bacteria, 114
    sudden increase in flow, 113
  rescue, 115
  rescue apparatus, 115
  safety at ground level, 114
  size of working parties, 115
Site managers, duties of, 153
Statistics
  ill-health, 10
  injuries, 3–6
  reliability of, 9–10
Statutory authorities, special requirements, 164
Structural framework, erection of, 75–86
  prevention of falls from, 76–82
  prevention of framework collapse, 83–86
    framework stability, 84
    precast concrete, 85–86
    trusses and girders, 84–85
  prevention of items from falling, 82–83
    small items, 83
    stacked materials, 83
  safe access, 76–78
    beam crossing, 77
    ladders, 76–77
  safe working places, 78–79
    safety belts and harnesses, 79–80
    safety nets, 80–81
    use of cranes, 81–82
Substances hazardous to health, 20

Telephone, site and portable, 219
Tort, 25–26
  definition, 25
  latent damage, 26
  limitation, 25–26
  negligence, 25
  negligence involving injury, 26
  vicarious liability, 27
Training, 204–207
  first aid, 207, 224–225
  graduates and technicians, 205
  induction courses, 204
  managers, 205–206
  resident engineers, 206
  safety advisers, 206
  safety supervisors, 207
Transport and mobile plant, 98–101
  site transport planning, 98–100
  ROPS and FOPS, 100–101
Tunnelling, 104–111
  compressed air working, 110–111
  hazards, 104–111
    atmospheric pollution, 106–108
    burial, 104–105
    electrical, 110
    falls, 106
    fire, 111
    inundation, 105–106
    noise, 109–110
    transport, 108–109
    trapping, 108
  safety audits, 111

Underground services, 103–104

Vicarious liability, 27

Water, work over, 118–121
  prevention of falls, 119
  rescue equipment, 119
  rescue facilities, 120
  water transport, 120–121